EINSTEIN'S ESSAYS IN SCIENCE

Albert Einstein

Translated from the German
by Alan Harris

Dover Publications, Inc.
Mineola, New York

Bibliographical Note

This Dover edition, first published in 2009, is an unabridged republication of *Essays in Science,* originally published in 1934 by The Wisdom Library (a division of The Philosophical Library, Inc.), New York.

Library of Congress Cataloging-in-Publication Data

Einstein, Albert, 1879-1955.
[Mein Weltbild. English]
Einstein's essays in science / Albert Einstein. — Dover ed.
p. cm.
Originally published: New York : Wisdom Library, 1934.
ISBN-13: 978-0-486-47011-5
ISBN-10: 0-486-47011-3
1. Physics. 2. Science — Philosophy 3. Peace. 4. Science — Social aspects. I. Title. II. Title: Essays in science.

QC71.E513 2009
530.092—dc22

2009000860

Manufactured in the United States of America
Dover Publications, Inc., 31 East 2nd Street, Mineola, N.Y. 11501

PREFACE TO ABRIDGED EDITION

The World as I See It in its original form includes essays by Einstein on Judaism, Germany, Politics and Pacifism and sundry other topics. These have been omitted in the present abridged edition. The object of this reprint is simply to give the general reader an opportunity to examine some of the distinguished scholar's papers dealing with science.

PUBLISHER'S NOTE

to the original edition, entitled
"The World as I See It"

Albert Einstein is one of the most modest people in the world. In the letters and papers in this book he constantly refers to his "accidental" fame and his "unearned popularity." So deep-rooted is this feeling that he has merely tilled his own row, howbeit that row was an inconceivably fertile one, that for many years his friends and intimates were unable to prevail upon him to publish his letters, papers and speeches, although opportunities and requests were showered on him. Even his scientific articles he would release to none save academic journals that do not reach the general public. The imagination of the world is stirred by the name of Albert Einstein, and yet hitherto it has had only the scantiest material to feed upon. Occasional, all-too-brief interviews; a few instances where Einstein suppressed his modesty for the sake of some great humanitarian cause; and the famous four-page pamphlet on the special theory of relativity.

The publishers do not know what good fortune at last prompted Albert Einstein to relax his attitude a little. Perhaps he felt that in times like these each man must sacrifice himself if he can help but a little to alleviate the horrible conditions which have fallen upon the world. If anything could break Einstein's silence, it was the threat of war in Europe. Perhaps, too, it was the persecution of his people, the Jews. Perhaps it was, in some small degree, a further extension of his modesty in a desire to show how much his scientific work has been prepared for, supported, and amplified by less heralded collaborators, as he explains in his articles and speeches on Kepler, Newton, Maxwell, Planck, Niels Bohr and others. Whatever the cause, he at last gave permission to one of his intimates, who prefers to conceal his identity beneath the initials J.H., because his was a labor of love, to collect and publish certain of his writings.

These papers were originally published by the Querido Verlag in Amsterdam. In fairness to Professor Einstein, his American publishers would like to make it clear that although they have his full authorization to translate the German text as published in Holland, and although the documents from which the original publication was made have his authentication, there has been no further collaboration by him. Because of this, the publishers have taken extra and unusual care to check every detail of the translation. An expert has gone over the work to make sure that it represents the exact meaning of Professor Einstein. Thus, the responsibility for the accuracy of the translation, though every attempt has been made to insure it, must not be placed with Professor Einstein.

There is a saying that only twelve people in the world can understand Einstein's theory of relativity. The difficulties of reaching such an understanding have heretofore been heightened by lack in print of anything but the abstract mathematical formulation of the theory. The papers and speeches published for the first time in this volume will be comprehensible to any well-educated person. They deal only in part with the central core of the theory; but the very fact that they carry its elaborations somewhat afield to particular applications and examples will help the scientific-minded layman to achieve a more complete comprehension of the theory itself. This book presents to the world at large for the first time what Einstein has really accomplished in the field of abstract physics. We cannot help but feel that its publication is an event of historical importance.

CONTENTS

PRINCIPLES OF RESEARCH

IN THE temple of Science are many mansions, and various indeed are they that dwell therein and the motives that have led them thither. Many take to science out of a joyful sense of superior intellectual power; science is their own special sport to which they look for vivid experience and the satisfaction of ambition; many others are to be found in the temple who have offered the products of their brains on this altar for purely utilitarian purposes. Were an angel of the Lord to come and drive all the people belonging to these two categories out of the temple, it would be noticeably emptier, but there would still be some men, of both present and past times, left inside. Our Planck is one of them, and that is why we love him.

I am quite aware that we have just now light-heartedly expelled in imagination many excellent men who are largely, perhaps chiefly, responsible for the building of the temple of Science; and in many cases our angel would find it a pretty ticklish job to decide. But of one thing I feel sure: if the types we have just expelled were the only types there were, the temple would never have existed, any more than one can have a wood consisting of nothing but creepers. For these people any sphere of human activity will do, if it comes to a point; whether they become officers, tradesmen or scientists depends on circumstances. Now let us have another look at those who have found favor with the angel. Most of them are somewhat odd, uncommunicative, solitary fellows, really less like each other, in spite of these common characteristics, than the hosts of the rejected. What has brought them to the temple? That is a difficult question and no single answer will cover it. To begin with I believe with Schopenhauer that one of the strongest motives that lead men to art and science is escape from everyday life with its painful crudity and hopeless dreariness, from the fetters of one's own ever shifting desires. A finely tempered nature longs to escape from personal life into the world of objective perception and thought; this desire may be compared with the townsman's irresistible longing to escape from his noisy, cramped surroundings into the silence of high mountains, where the eye ranges freely through the still, pure air and fondly traces out the restful contours apparently built for eternity. With this negative motive there goes a personal one. Man tries to make for himself in the fashion that suits him best a simplified and intelligible picture of the world; he then tries to some extent to substitute this cosmos of his for the world of experience, and thus to overcome it. This is what the painter, the poet, the speculative philosopher and the natural scientist do, each in his own fashion. He makes this cosmos and its construction the pivot of his emotional life, in order to find in this way the peace and security which he cannot find in the narrow whirlpool of personal experience.

What place does the theoretical physicist's picture of the world occupy among all these possible pictures? It demands the highest possible standard of rigorous precision in the description of relations, such as only the use of mathematical language can give. In regard to his subject matter, on the other hand, the physicist has to limit himself very severely: he must content himself with describing the most simple events which can be brought within the domain of our experience; all events of a more complex order are beyond the power of the human intellect to reconstruct with the subtle accuracy and logical perfection which the theoretical physicist demands. Supreme purity, clarity and certainty at the cost of completeness. But what can be the attraction of getting to know such a tiny section of nature thoroughly, while one leaves everything subtler and more complex shyly and timidly alone? Does the product of such a modest effort deserve to be called by the proud name of a theory of the Universe?

In my belief the name is justified; for the general laws on which the structure of theoretical physics is based claim to be valid for any natural phenomenon whatsoever. With them, it ought to be possible to arrive at the description, that is to say, the theory, of every natural process, including life, by means of pure deduction, if that process of deduction were not far beyond the capacity of the human intellect. The physicist's renunciation of completeness for his cosmos is therefore not a matter of fundamental principle.

The supreme task of the physicist is to arrive at those universal elementary laws from which the cosmos can be built up by pure deduction. There is no logical path to these laws; only intuition, resting on sympathetic understanding of experience, can reach them. In this methodological uncertainty, one might suppose that there were any number of possible systems of theoretical physics all with an equal amount to be said for them; and this opinion is no doubt correct, theoretically. But evolution has shown that at any given moment, out of all conceivable constructions, a single one has always proved itself absolutely superior to all the rest. Nobody who has really gone deeply into the matter will deny that in practice the world of phenomena uniquely determines the theoretical system, in spite of the fact that there is no logical bridge between phenomena and their theoretical principles; this is what Leibnitz described so happily as a "pre-established harmony." Physicists often accuse epistemologists of not paying sufficient attention to this fact. Here, it seems to me, lie the roots of the controversy carried on some years ago between Mach and Planck.

The longing to behold this pre-established harmony is the source of the inexhaustible patience and endurance with which Planck has devoted himself, as we see, to the most general problems of our science, refusing to let himself be diverted to more grateful and more easily attained ends. I have often heard colleagues try to attribute this attitude of his to extraordinary will-power and discipline—wrongly, in my opinion. The state of mind which enables a man to do work of this kind is akin to that of the religious worshiper or the lover; the daily effort comes from no deliberate intention or program, but straight from the heart. There he sits, our beloved Planck, and smiles inside himself at my childish playing-about with the lantern of Diogenes. Our affection for him needs no threadbare explanation. May the love of science continue to illumine his path in the future and lead him to the solution of the most important problem in present-day physics, which he has himself posed and done so much to solve. May he succeed in uniting the quantum theory and electrodynamics in a single logical system.

(Address on the occasion of Max Planck's sixtieth birthday delivered at the Physical Society in Berlin)

INAUGURAL ADDRESS TO THE PRUSSIAN ACADEMY OF SCIENCES (1914)

Gentlemen,

First of all, I have to thank you most heartily for conferring the greatest benefit on me that anybody can confer on a man like myself. By electing me to your Academy you have freed me from the distractions and cares of a professional life and so made it possible for me to devote myself entirely to scientific studies. I beg that you will continue to believe in my gratitude and my industry even when my efforts seem to you to yield but a poor result.

Perhaps I may be allowed a propos of this to make a few general remarks on the relation of my sphere of activity, which is theoretical physics, towards experimental physics. A mathematician friend of mine said to me the other day half in jest: "The mathematician can do a lot of things, but never what you happen to want him to do just at the moment." Much the same often applies to the theoretical physicist when the experimental physicist calls him in. What is the reason for this peculiar lack of adaptability?

The theorist's method involves his using as his foundation general postulates or "principles" from which he can deduce conclusions. His work thus falls into two parts. He must first discover his principles and then draw the conclusions which follow from them. For the second of these tasks he receives an admirable equipment at school. Once, therefore, he has performed the first task in some department, or for some complex of related phenomena, he is certain of success, provided his industry and intelligence are adequate. The first of these tasks, namely, that of establishing the principles which are to serve as the starting point of his deduction, is of an entirely different nature. Here there is no method capable of being learned and systematically applied so that it leads to the goal. The scientist has to worm these general principles out of nature by perceiving certain general features which permit of precise formulation, amidst large complexes of empirical facts.

Once this formulation is successfully accomplished, inference follows on inference, often revealing relations which extend far beyond the province of the reality from which the principles were drawn. But as long as the principles capable of serving as starting points for the deduction remain undiscovered, the individual fact is of no use to the theorist; indeed he cannot even do anything with isolated empirical generalizations of more or less wide application. No, he has to persist in his helpless attitude towards the separate results of empirical research, until principles which he can make the basis of deductive reasoning have revealed themselves to him.

This is the kind of position in which theory finds

itself at present in regard to the laws of heat, radiation, and molecular movement at low temperatures. About fifteen years ago nobody had yet doubted that a correct account of the electrical, optical and thermal properties of bodies was possible on the basis of Galileo-Newtonian mechanics applied to the movement of molecules and of Clerk Maxwell's theory of the electro-magnetic field. Then Planck showed that in order to establish a law of heat radiation consonant with experience, it was necessary to employ a method of calculation the incompatibility of which with the principles of classical physics became clearer and clearer. For with this method of calculation Planck introduced the quantum hypothesis into physics, which has since received brilliant confirmation. With this quantum hypothesis he dethroned classical physics as applied to the case where sufficiently small masses are moved at sufficiently low speeds and high rates of acceleration, so that today the laws of motion propounded by Galileo and Newton can only be allowed validity as limiting laws. In spite of assiduous efforts, however, the theorists have not yet succeeded in replacing the principles of mechanics by others which fit in with Planck's law of heat radiation or the quantum hypothesis. No matter how definitely it has been proved that heat is to be explained by molecular movement, we have nevertheless to admit today that our position in regard to the fundamental laws of this motion resembles that of astronomers before Newton in regard to the motions of the planets.

I have just now referred to a group of facts for the theoretical treatment of which the principles are lacking. But it may equally well happen that clearly formulated principles lead to conclusions which fall entirely, or almost entirely, outside the sphere of reality at present accessible to our experience. In that case it may need many years of empirical research to ascertain whether the theoretical principles correspond with reality. We have an instance of this in the theory of relativity.

An analysis of the fundamental concepts of space and time has shown us that the principle of the constant velocity of light in empty space, which emerges from the optics of bodies in motion, by no means forces us to accept the theory of a stationary luminiferous ether. On the contrary, there is nothing to prevent our framing a general theory which takes account of the fact that in experiments carried out on the earth we are wholly unconscious of the translatory motion of the earth. This involves using the principle of relativity, which says that the laws of nature do not alter their form when one proceeds from the original (legitimate) system of co-ordinates to a new one which is in uniform translatory motion with respect to it. This theory has received impressive confirmation from experience and has led to a simplification of the theoretical description of groups of facts already connected together.

On the other hand, from the theoretical point of view this theory is not wholly satisfactory, because the principle of relativity just formulated prefers *uniform* motion. If it is true that no absolute significance can be attached to *uniform* motion from the physical point of view, the question arises whether this statement must not also be extended to non-uniform motions. It became clear that one arrives at a quite definite enlargement of the relativity theory if one postulates a principle of relativity in this extended sense. One is led thereby to a general theory of gravitation which includes dynamics. For the present, however, we have not the necessary array of facts to test the legitimacy of our introduction of the postulated principle.

We have ascertained that inductive physics asks questions of deductive, and vice versa, to answer which demands the exertion of all our energies. May we soon succeed in making permanent progress by our united efforts!

ON SCIENTIFIC TRUTH

(1) It is difficult even to attach a precise meaning to the term "scientific truth." So different is the meaning of the word "truth" according to whether we are dealing with a fact of experience, a mathematical proposition or a scientific theory. "Religious truth" conveys nothing clear to me at all.
(2) Scientific research can reduce superstition by encouraging people to think and survey things in terms of cause and effect. Certain it is that a conviction, akin to religious feeling, of the rationality or intelligibility of the world lies behind all scientific work of a higher order.
(3) This firm belief, a belief bound up with deep feeling, in a superior mind that reveals itself in the world of experience, represents my conception of God. In common parlance this may be described as "pantheistic" (Spinoza).
(4) Denominational traditions I can only consider historically and psychologically; they have no other significance for me.

ON THE METHOD OF THEORETICAL PHYSICS

IF YOU want to find out anything from the theoretical physicists about the methods they use, I advise you to stick closely to one principle: don't listen to their words, fix your attention on their deeds. To him who is a discoverer in this field the products of his imagination appear so necessary and natural that he regards them, and would like to have them regarded by others, not as creations of thought but as given realities.

These words sound like an invitation to you to walk out of this lecture. You will say to yourselves, the fellow's a working physicist himself and ought therefore to leave all questions of the structure of theoretical science to the epistemologists.

Against such criticism I can defend myself from the personal point of view by assuring you that it is not at my own instance but at the kind invitation of others that I have mounted this rostrum, which serves to commemorate a man who fought hard all his life for the unity of knowledge. Objectively, however, my enterprise can be justified on the ground that it may, after all, be of interest to know how one who has spent a life-time in striving with all his might to clear up and rectify its fundamentals looks upon his own branch of science. The way in which he regards its past and present may depend too much on what he hopes for the future and aims at in the present; but that is the inevitable fate of anybody who has occupied himself intensively with a world of ideas. The same thing happens to him as to the historian, who in the same way, even though perhaps unconsciously, groups actual events around ideals which he has formed for himself on the subject of human society.

Let us now cast an eye over the development of the theoretical system, paying special attention to the relations between the content of the theory and the totality of empirical fact. We are concerned with the eternal antithesis between the two inseparable components of our knowledge, the empirical and the rational, in our department.

We reverence ancient Greece as the cradle of western science. Here for the first time the world witnessed the miracle of a logical system which proceeded from step to step with such precision that every single one of its propositions was absolutely indubitable—I refer to Euclid's geometry. This admirable triumph of reasoning gave the human intellect the necessary confidence in itself for its subsequent achievements. If Euclid failed to kindle your youthful enthusiasm, then you were not born to be a scientific thinker.

But before mankind could be ripe for a science which takes in the whole of reality, a second fundamental truth was needed, which only became common property among philosophers with the advent of Kepler and Galileo. Pure logical thinking cannot yield us any knowledge of the empirical world; all knowledge of reality starts from experience and ends in it. Propositions arrived at by purely logical means are completely empty as regards reality. Because Galileo saw this, and particularly because he drummed it into the scientific world, he is the father of modern physics—indeed, of modern science altogether.

If, then, experience is the alpha and the omega of all our knowledge of reality, what is the function of pure reason in science?

A complete system of theoretical physics is made up of concepts, fundamental laws which are supposed to be valid for those concepts and conclusions to be reached by logical deduction. It is these conclusions which must correspond with our separate experiences; in any theoretical treatise their logical deduction occupies almost the whole book.

This is exactly what happens in Euclid's geometry, except that there the fundamental laws are called axioms and there is no question of the conclusions having to correspond to any sort of experience. If, however, one regard Euclidean geometry as the science of the possible mutual relations of practically rigid bodies in space, that is to say, treats it as a physical science, without abstracting from its original empirical content, the logical homogeneity of geometry and theoretical physics becomes complete.

We have thus assigned to pure reason and experience their places in a theoretical system of physics. The structure of the system is the work of reason; the empirical contents and their mutual relations must find their representation in the conclusions of the theory. In the possibility of such a representation lie the sole value and justification of the whole system, and especially of the concepts and fundamental principles which underlie it. These latter, by the way, are free inventions of the human intellect, which cannot be justified either by the nature of that intellect or in any other fashion *a priori.*

These fundamental concepts and postulates, which cannot be further reduced logically, form the essential part of a theory, which reason cannot touch. It is the grand object of all theory to make these irreducible elements as simple and as few in number as possible, without having to renounce the adequate representation of any empirical content whatever.

The view I have just outlined of the purely fictitious character of the fundamentals of scientific theory was by no means the prevailing one in the eighteenth or even the nineteenth century. But it is steadily gaining ground from the fact that the distance in thought between the fundamental concepts and laws on one side and, on the other, the conclusions which have to be brought into relation with our experience grows larger and larger, the simpler the logical structure becomes—that is to say, the smaller the number of logically independent conceptual elements which are found necessary to support the structure.

Newton, the first creator of a comprehensive,

workable system of theoretical physics, still believed that the basic concepts and laws of his system could be derived from experience. This is no doubt the meaning of his saying, *hypotheses non fingo*.

Actually the concepts of time and space appeared at that time to present no difficulties. The concepts of mass, inertia and force, and the laws connecting them seemed to be drawn directly from experience. Once this basis is accepted, the expression for the force of gravitation appears derivable from experience, and it was reasonable to hope for the same in regard to other forces.

We can indeed see from Newton's formulation of it that the concept of absolute space, which comprised that of absolute rest, made him feel uncomfortable; he realized that there seemed to be nothing in experience corresponding to this last concept. He was also not quite comfortable about the introduction of forces operating at a distance. But the tremendous practical success of his doctrines may well have prevented him and the physicists of the eighteenth and nineteenth centuries from recognizing the fictitious character of the foundations of his system.

The natural philosophers of those days were, on the contrary, most of them possessed with the idea that the fundamental concepts and postulates of physics were not in the logical sense free inventions of the human mind but could be deduced from experience by "abstraction"—that is to say by logical means. A clear recognition of the erroneousness of this notion really only came with the general theory of relativity, which showed that one could take account of a wider range of empirical facts, and that too in a more satisfactory and complete manner, on a foundation quite different from the Newtonian. But quite apart from the question of the superiority of one or the other, the fictitious character of fundamental principles is perfectly evident from the fact that we can point to two essentially different principles, both of which correspond with experience to a large extent; this proves at the same time that every attempt at a logical deduction of the basic concepts and postulates of mechanics from elementary experiences is doomed to failure.

If, then, it is true that this axiomatic basis of theoretical physics cannot be extracted from experience but must be freely invented, can we ever hope to find the right way? Nay more, has this right way any existence outside our illusions? Can we hope to be guided in the right way by experience when there exist theories (such as classical mechanics) which to a large extent do justice to experience, without getting to the root of the matter? I answer without hesitation that there is, in my opinion, a right way, and that we are capable of finding it. Our experience hitherto justifies us in believing that nature is the realization of the simplest conceivable mathematical ideas. I am convinced that we can discover by means of purely mathematical constructions the concepts and the laws connecting them with each other, which furnish the key to the understanding of natural phenomena. Experience may suggest the appropriate mathematical concepts, but they most certainly cannot be deduced from it. Experience remains, of course, the sole criterion of the physical utility of a mathematical construction. But the creative principle resides in mathematics. In a certain sense, therefore, I hold it true that pure thought can grasp reality, as the ancients dreamed.

In order to justify this confidence, I am compelled to make use of a mathematical conception. The physical world is represented as a four-dimensional continuum. If I assume a Riemannian metric in it and ask what are the simplest laws which such a metric system can satisfy, I arrive at the relativist theory of gravitation in empty space. If in that space I assume a vector-field or an anti-symmetrical tensor-field which can be inferred from it, and ask what are the simplest laws which such a field can satisfy, I arrive at Clerk Maxwell's equations for empty space.

At this point we still lack a theory for those parts of space in which electrical density does not disappear. De Broglie conjectured the existence of a wave field. which served to explain certain quantum properties of matter. Dirac found in the spinors field-magnitudes of a new sort, whose simplest equations enable one to a large extent to deduce the properties of the electron. Subsequently I discovered, in conjunction with my colleague, that these spinors form a special case of a new sort of field, mathematically connected with the four-dimensional system, which we called "semivectors." The simplest equations to which such semivectors can be reduced furnish a key to the understanding of the existence of two sorts of elementary particles, of different ponderable mass and equal but opposite electrical charge. These semivectors are, after ordinary vectors, the simplest mathematical fields that are possible in a metrical continuum of four dimensions, and it looks as if they described, in an easy manner, certain essential properties of electrical particles.

The important point for us to observe is that all these constructions and the laws connecting them can be arrived at by the principle of looking for the mathematically simplest concepts and the link between them. In the limited nature of the mathematically existent simple fields and the simple equations possible between them, lies the theorist's hope of grasping the real in all its depth.

Meanwhile the great stumbling-block for a field-theory of this kind lies in the conception of the atomic structure of matter and energy. For the theory is fundamentally non-atomic in so far as it operates exclusively with continuous functions of space, in contrast to classical mechanics, whose most important element, the material point, in itself does justice to the atomic structure of matter.

The modern quantum theory in the form associated with the names of de Broglie, Schrödinger, and Dirac, which operates with continuous functions, has overcome these difficulties by a bold piece of interpretation which was first given a clear form by Max Born. According to this, the spatial functions which appear in the equations make no claim to be a mathematical model of the atomic structure. Those functions are only supposed to determine the mathematical probabilities of the occurrence of such structures if measurements were taken at a particular spot or in a certain state of motion. This notion is logically unobjectionable and has important successes to its credit. Unfortunately, however, it compels one to use a continuum the number of whose dimensions is not that ascribed to space by physics hitherto (four) but rises indefinitely with the number of the particles constituting the system under consideration. I cannot but confess that I attach only a transitory importance to this interpretation. I still believe in the possibility of a model of reality—that is to say, of a theory which represents things themselves and not merely the probability of their occurrence.

On the other hand it seems to me certain that we must give up the idea of a complete localization of the particles in a theoretical model. This seems to me to be the permanent upshot of Heisenberg's principle of uncertainty. But an atomic theory in the true sense of the word (not merely on the basis of an interpretation) without localization of particles in a mathematical model, is perfectly thinkable. For instance, to account for the atomic character of electricity, the field equations need only lead to the following conclusions: A portion of space (three-dimensional) at whose boundaries electrical density disappears everywhere, always contains a total electrical charge whose size is represented by a whole number. In a continuum-theory atomic characteristics would be satisfactorily expressed by integral laws without localization of the formation entity which constitutes the atomic structure.

Not until the atomic structure has been successfully represented in such a manner would I consider the quantum-riddle solved.

JOHANNES KEPLER

IN ANXIOUS and uncertain times like ours, when it is difficult to find pleasure in humanity and the course of human affairs, it is particularly consoling to think of the serene greatness of a Kepler. Kepler lived in an age in which the reign of law in nature was by no means an accepted certainty. How great must his faith in a uniform law have been, to have given him the strength to devote ten years of hard and patient work to the empirical investigation of the movement of the planets and the mathematical laws of that movement, entirely on his own, supported by no one and understood by very few! If we would honor his memory worthily, we must get as clear a picture as we can of his problem and the stages of its solution.

Copernicus had opened the eyes of the most intelligent to the fact that the best way to get a clear grasp of the apparent movements of the planets in the heavens was by regarding them as movements around the sun conceived as stationary. If the planets moved uniformly in a circle around the sun, it would have been comparatively easy to discover how these movements must look from the earth. Since, however, the phenomena to be dealt with were much more complicated than that, the task was a far harder one.

The first thing to be done was to determine these movements empirically from the observations of Tycho Brahe. Only then did it become possible to think about discovering the general laws which these movements satisfy.

To grasp how difficult a business it was even to find out about the actual rotating movements, one has to realize the following. One can never see where a planet really is at any given moment, but only in what direction it can be seen just then from the earth, which is itself moving in an unknown manner around the sun. The difficulties thus seemed practically unsurmountable.

Kepler had to discover a way of bringing order into this chaos. To start with, he saw that it was necessary first to try and find out about the motion of the earth itself. This would simply have been impossible if there had existed only the sun, the earth and the fixed stars, but no other planets. For in that case one could ascertain nothing empirically except how the direction of the straight sun-earth line changes in the course of the year (apparent movement of the sun with reference to the fixed stars). In this way it was possible to discover that these sun-earth directions all lay in a plane stationary with reference to the fixed stars, at least according to the accuracy of observation achieved in those days, when there were no telescopes. By this means it could also be ascertained in what manner the line sun-earth revolves round the sun. It turned out that the angular velocity of this motion went through a regular change in the course of the year. But this was not of much use, as it was still not known how the distance from the earth to the sun alters in the course of the year. It was only when they found out about these changes that the real shape of the earth's orbit and the manner in which it is described were discovered.

Kepler found a marvelous way out of this dilemma. To begin with it was apparent from observations of the sun that the apparent path of the sun against the background of the fixed stars differed in speed at different times of the year, but that the angular velocity of this movement was always the same at the same point in the astronomical year, and therefore that the speed of rotation of the straight line earth-sun was always the same when it pointed to the same region of the fixed stars. It was thus legitimate to suppose that the earth's orbit was a self-enclosed one, described by the earth in the same way every year—which was by no means obvious *a priori*. For the adherent of the Copernican system it was thus as good as certain that this must also apply to the orbits of the rest of the planets.

This certainty made things easier. But how to ascertain the real shape of the earth's orbit? Imagine a brightly shining lantern M somewhere in the plane of the orbit. We know that this lantern remains permanently in its place and thus forms a kind of fixed triangulation point for determining the earth's orbit, a point which the inhabitants of the earth can take a sight on at any time of year. Let this lantern M be further away from the sun than the earth. With the help of such a lantern it was possible to determine the earth's orbit, in the following way:—

First of all, in every year there comes a moment when the earth E lies exactly on the line joining the sun S and the lantern M. If at this moment we look from the earth E at the lantern M, our line of sight will coincide with the line SM (sun-lantern). Suppose the latter to be marked in the heavens. Now imagine the earth in a different position and at a different time. Since the sun S and the lantern M can both be seen from the Earth, the angle at E in the triangle SEM is known. But we also know the direction of SE in relation to the fixed stars through direct solar observations, while the direction of the line SM in relation to the fixed stars was finally ascertained previously. But in the triangle SEM we also know the angle at S. Therefore, with the base SM arbitrarily laid down on a sheet of paper, we can, in virtue of our knowledge of the angles at E and S, construct the triangle SEM. We might do this at frequent intervals during the year; each time we should get on our piece of paper a position of the earth E with a date attached to it and a certain position in relation to the permanently fixed base SM. The earth's orbit would thereby be empirically determined, apart from its absolute size, of course.

But, you will say, where did Kepler get his lantern M? His genius and Nature, benevolent in this case, gave it to him. There was, for example, the planet Mars; and the length of the Martian year—i.e., one rotation of Mars around the sun—was known. It might happen one fine day that the sun, the earth and Mars lie absolutely in the same straight line. This position of Mars regularly recurs after one, two, etc., Martian years, as Mars has a self-enclosed orbit. At these known moments, therefore, SM always presents the same base, while the earth is always at a different point in its orbit. The observations of the sun and Mars at these moments thus constitute a means of determining the true orbit of the earth, as Mars then plays the part of our imaginary lantern. Thus it was that Kepler discovered the true shape of the earth's orbit and the way in which the earth describes it, and we who come after—Europeans, Germans, or even Swabians, may well admire and honor him for it.

Now that the earth's orbit had been empirically determined, the true position and length of the line SE at any moment was known, and it was not so terribly difficult for Kepler to calculate the orbits and motions of the rest of the planets too from observations—at least in principle. It was nevertheless an immense work, especially considering the state of mathematics at the time.

Now came the second and no less arduous part of Kepler's life work. The orbits were empirically known, but their laws had to be deduced from the empirical data. First he had to make a guess at the mathematical nature of the curve described by the orbit, and then try it out on a vast assemblage of figures. If it did not fit, another hypothesis had to be devised and again tested. After tremendous search, the conjecture that the orbit was an ellipse with the sun at one of its foci was found to fit the facts. Kepler also discovered the law governing the variation in speed during rotation, which is that the line sun-planet sweeps out equal areas in equal periods of time. Finally he also discovered that the square of the period of circulation around the sun varies as the cube of the major axes of the ellipse.

Our admiration for this splendid man is accompanied by another feeling of admiration and reverence, the object of which is no man but the mysterious harmony of nature into which we are born. As far back as ancient times people devised the lines exhibiting the simplest conceivable form of regularity. Among these, next to the straight line and the circle, the most important were the ellipse and the hyperbola. We see the last two embodied—at least very nearly so—in the orbits of the heavenly bodies.

It seems that the human mind has first to construct forms independently before we can find them in things. Kepler's marvelous achievement is a particularly fine example of the truth that knowledge cannot spring from experience alone but only from the comparison of the inventions of the intellect with observed fact.

THE MECHANICS OF NEWTON AND THEIR INFLUENCE ON THE DEVELOPMENT OF THEORETICAL PHYSICS

It is just two hundred years ago that Newton closed his eyes. It behooves us at such a moment to remember this brilliant genius, who determined the course of western thought, research and practice to an extent that nobody before or since his time can touch. Not only was he brilliant as an inventor of certain key methods, but he also had a unique command of the empirical material available in his day, and he was marvelously inventive as regards mathematical and physical methods of proof in individual cases. For all these reasons he deserves our deepest reverence. The figure of Newton has, however, an even greater importance than his genius warrants because of the fact that destiny placed him at a turning point in the history of the human intellect. To see this vividly, we have to remind ourselves that before Newton there existed no self-contained system of physical causality which was capable of representing any of the deeper features of the empirical world.

No doubt the great materialists of ancient Greece had insisted that all material events should be traced back to a strictly regular series of atomic movements, without admitting any living creature's will as an independent cause. And no doubt Descartes had in his own way taken up this quest again. But it remained a bold ambition, the problematical ideal of a school of philosophers. Actual results of a kind to support the belief in the existence of a complete chain of physical causation hardly existed before Newton.

Newton's object was to answer the question: Is there such a thing as a simple rule by which one can calculate the movements of the heavenly bodies in our planetary system completely, when the state of motion of all these bodies at one moment is known? Kepler's empirical laws of planetary movement, deduced from Tycho Brahe's observations, confronted him, and demanded explanation.[1] These laws gave, it is true, a complete answer to the question of *how* the planets move around the sun (the elliptical shape of the orbit, the sweeping of equal areas by the radii in equal times, the relation between the major axes and the period of circulation around the sun); but they did not satisfy the demand for causal explanation. They are three logically independent rules, revealing no inner connection with each other. The third law cannot simply be transferred quantitatively to other central bodies than the sun (there is, e.g., no relation

[1] Today everybody knows what prodigious industry was needed to discover these laws from the empirically ascertained orbits. But few pause to reflect on the brilliant methods by which Kepler deduced the real orbits from the apparent ones—i.e., from the movements as they were observed from the earth.

between the rotatory period of a planet around the sun and that of a moon around its planet). The most important point, however, is this: these laws are concerned with the movement as a whole, and not with the question *how the state of motion of a system gives rise to that which immediately follows it in time;* they are, as we should say now, integral and not differential laws.

The differential law is the only form which completely satisfies the modern physicist's demand for causality. The clear conception of the differential law is one of Newton's greatest intellectual achievements. It was not merely this conception that was needed but also a mathematical formalism, which existed in a rudimentary form but needed to acquire a systematic form. Newton found this also in the differential and the integral calculus. We need not consider the question here whether Newton hit upon the same mathematical methods independently of Leibnitz or not. In any case it was absolutely necessary for Newton to perfect them, since they alone could provide him with the means of expressing his ideas.

Galileo had already moved a considerable way towards a knowledge of the law of motion. He discovered the law of inertia and the law of bodies falling freely in the gravitational field of the earth, namely, that a mass (more accurately, a mass-point) which is unaffected by other masses moves uniformly and in a straight line. The vertical speed of a free body in the gravitational field increases uniformly with the time. It may seem to us today to be but a short step from Galileo's discoveries to Newton's law of motion. But it should be observed that both the above statements refer in their form to the motion as a whole, while Newton's law of motion provides an answer to the question: how does the state of motion of a mass-point behave in an *infinitely short time* under the influence of an external force? It was only by considering what takes place during an infinitely short time (the differential law) that Newton reached a formula which applies to all motion whatsoever. He took the conception of force from the science of statics which had already reached a high stage of development. The connection of force and acceleration was only made possible for him by the introduction of the new concept of mass, which was supported, strange to say, by an illusory definition. We are so accustomed today to the creation of concepts corresponding to differential quotients that we can now hardly grasp any longer what a remarkable power of abstraction it needed to reach the general differential law by a double crossing of frontiers, in the course of which the concept of mass had in addition to be invented.

But a causal conception of motion was still far from being achieved. For the motion was only determined by the equation of motion in cases where the force was given. Inspired no doubt by the uniformity of planetary motions, Newton conceived the idea that the force operating on a mass was determined by the position of all masses situated at a sufficiently small distance from the mass in question. It was not till this connection was established that a completely causal conception of motion was achieved. How Newton, starting from Kepler's laws of planetary motion, performed this task for gravitation and so discovered that the kinetic forces acting on the stars and gravity were of the same nature, is well known. It is the combination of the Law of Motion with the Law of Attraction which constitutes that marvelous edifice of thought which makes it possible to calculate the past and future states of a system from the state obtaining at one particular moment, in so far as the events take place under the influence of the forces of gravity alone. The logical completeness of Newton's conceptual system lay in this, that the only things that figure as causes of the acceleration of the masses of a system are *these masses themselves.*

On the strength of the foundation here briefly sketched Newton succeeded in explaining the motions of the planets, moons and comets down to the smallest details, as well as the tides and the precessional movement of the earth—a deductive achievement of unique magnificence. The discovery that the cause of the motions of the heavenly bodies is identical with the gravity with which we are so familiar from everyday life must have been particularly impressive.

But the importance of Newton's achievement was not confined to the fact that it created a workable and logically satisfactory basis for the actual science of mechanics; up to the end of the nineteenth century it formed the program of every worker in the field of theoretical physics. All physical events were to be traced back to masses subject to Newton's laws of motion. The law of force simply had to be widened and adapted to the type of event under consideration. Newton himself tried to apply this scheme to optics, assuming light to consist of inert corpuscles. Even the wave theory of light made use of Newton's law of motion, after it had been applied to the mass of a continuum. Newton's equations of motion were the sole basis of the kinetic theory of heat, which not only prepared people's minds for the discovery of the law of the conservation of energy but also led to a theory of gases which has been confirmed down to the last detail, and a more profound view of the nature of the second law of thermodynamics. The development of electricity and magnetism has proceeded right down to our own day along Newtonian lines (electrical and magnetic substance, forces acting at a distance). Even the revolution in electrodynamics and optics brought about by Faraday and Clerk Maxwell, which formed the first great fundamental advance in theoretical physics since Newton, took place entirely under the ægis of Newton's ideas. Clerk Maxwell, Boltzmann, and Lord Kelvin never wearied of tracing the electromagnetic fields and their reciprocal dynamic actions back to the mechanical action of hypothetical continuously diffused masses. As a result, however, of the hopelessness or at any rate the lack of success of those efforts, a gradual revolution in our fundamental notions has taken place since the end of the nineteenth century; theoretical physics have outgrown the Newtonian frame which gave stability and intellectual guidance to science for nearly two hundred years.

Newton's fundamental principles were so satisfactory from the logical point of view that the impetus to overhaul them could only spring from the imperious demands of empirical fact. Before I go into this I must insist that Newton himself was better aware of the weaknesses inherent in his intellectual edifice than the generations of scientists which followed him. This fact has always roused my respectful admiration, and I should like therefore to dwell on it for a moment.

I. In spite of the fact that Newton's ambition to represent his system as necessarily conditioned by experience and to introduce the smallest possible number of concepts not directly referable to empirical objects is everywhere evident, he sets up the concept of absolute space and absolute time, for which he has often been criticized in recent years. But in this point Newton is particularly consistent. He had realized that observable geometrical magnitudes (distances of material points from one another) and their course in time do not completely characterize motion in its physical aspects. He proved this in the famous experiment with the rotating vessel of water. Therefore, in addition to masses and temporally variable distances, there must be something else that determines motion. That "something" he takes to be relation to "absolute space." He is aware that space must possess a kind of physical reality if his laws of motion are to have any meaning, a reality of the same sort as material points and the intervals between them.

The clear realization of this reveals both Newton's wisdom and also a weak side to his theory. For the logical structure of the latter would undoubtedly be more satisfactory without this shadowy concept; in that case only things whose relations to perception are perfectly clear (mass-points, distances) would enter into the laws.

II. The introduction of forces acting directly and instantaneously at a distance into the representation of the effects of gravity is not in keeping with the character of most of the processes familiar to us from everyday life. Newton meets this objection by pointing to the fact that his law of reciprocal gravitation is not supposed to be a final explanation but a rule derived by induction from experience.

III. Newton's teaching provided no explanation for the highly remarkable fact that the weight and the inertia of a body are determined by the same quantity (its mass). The remarkableness of this fact struck Newton himself.

None of these three points can rank as a logical objection to the theory. In a sense they merely represent unsatisfied desires of the scientific spirit in its struggle for a complete and unitary penetration of natural events by thought.

Newton's doctrine of motion, considered as the key idea of the whole of theoretical physics, received its first shock from Clerk Maxwell's theory of electricity. It became clear that the reciprocal actions between bodies due to electric and magnetic forces were affected, not by forces operating instantaneously at a distance, but by processes which are propagated through space at a finite speed. Faraday conceived a new sort of real physical entity, namely the "field," in addition to the mass-point and its motion. At first people tried, clinging to the mechanical mode of thought, to look upon it as a mechanical condition (motion or force) of a hypothetical medium by which space was filled up (the ether). But when this interpretation refused to work in spite of the most obstinate efforts, people gradually got used to the idea of regarding the "electro-magnetic field" as the final irreducible constituent of physical reality. We have H. Hertz to thank for definitely freeing the conception of the field from all encumbrances derived from the conceptual armory of mechanics, and H. A. Lorentz for freeing it from a material substratum; according to the latter the only thing left to act as a substratum for the field was physical, empty space (or ether), which even in the mechanics of Newton had not been destitute of all physical functions. By the time this point was reached, nobody any longer believed in immediate momentary action at a distance, not even in the sphere of gravitation, even though no field theory of the latter had been clearly sketched out owing to lack of sufficient factual knowledge. The development of the theory of the electro-magnetic field—once Newton's hypothesis of forces acting at a distance had been abandoned—led to the attempt to explain the Newtonian law of motion on electro-magnetic lines or alternatively to replace it by a more accurate one based on the field-theory. Even if these efforts did not meet with complete success, still the fundamental concepts of mechanics had ceased to be looked upon as fundamental constituents of the physical cosmos.

The theory of Clerk Maxwell and Lorentz led inevitably to the special theory of relativity, which ruled out the existence of forces acting at a distance, and resulted in the destruction of the notion of absolute simultaneity. This theory made it clear that mass is not a constant quantity but depends on (indeed it is equivalent to) the amount of energy content. It also showed that Newton's law of motion was only to be regarded as a limiting law valid for small velocities; in its place it set up a new law of motion in which the speed of light in vacuo figures as the critical velocity.

The general theory of relativity formed the last step in the development of the program of the field-theory. Quantitatively it modified Newton's theory only slightly, but for that all the more profoundly qualitatively. Inertia, gravitation, and the metrical behavior of bodies and clocks were reduced to a single field quality; this field itself was again placed in dependence on bodies (generalization of Newton's law of gravity or the field law corresponding to it, as formulated by Poisson). Space and time were thereby divested not of their reality but of their causal absoluteness (absoluteness affecting but not affected) which Newton had been compelled to ascribe to them in order to be able to give expression to the laws then known. The generalized law of inertia takes over the function of Newton's law of motion. This short account is enough to show how the elements of Newtonian theory passed over into the general theory of relativity, whereby the three defects above mentioned were overcome. It looks as if the law of motion could be deduced from the field law corresponding to the Newtonian law of force. Only when this goal has been completely reached will it be possible to talk about a pure field-theory.

In a more formal sense also Newton's mechanics prepared the way for the field-theory. The application of Newton's mechanics to continuously distributed masses led inevitably to the discovery and application of partial differential equations, which in their turn first provided the language for the laws of the field-theory. In this formal respect Newton's conception of the differential law constitutes the first decisive step in the development which followed.

The whole evolution of our ideas about the processes of nature, with which we have been concerned so far, might be regarded as an organic development of Newton's ideas. But while the process of perfecting the field-theory was still in full swing, the facts of heat-radiation, the spectra, radio-activity, etc., revealed a limit to the serviceableness of the whole intellectual system which today still seems to us absolutely insuperable in spite of immense successes at certain points. Many physicists maintain—and there are weighty arguments in their favor—that in the face of these facts not merely the differential law but the law of causation itself—hitherto the fundamental postulate of all natural science—has collapsed. Even the possibility of a spatio-temporal construction, which can be unambiguously coordinated with physical events, is denied. That a mechanical system is permanently susceptible only of discrete energy-values or states—as experience so to speak directly shows—seems at first sight hardly deducible from a field-theory which operates with differential equations. The de Broglie-Schrödinger method, which has in a certain sense the character of a field-theory, does indeed deduce the existence of only-discrete states, in astonishing agreement with empirical fact, on a basis of differential equations operating with a kind of resonance-theory, but it has to do without a localization of the mass-particles and without strictly causal laws. Who would presume today to decide the question whether the law of causation and the differential law, these ultimate premises of the Newtonian view of nature, must definitely be given up?

CLERK MAXWELL'S INFLUENCE ON THE EVOLUTION OF THE IDEA OF PHYSICAL REALITY

THE BELIEF in an external world independent of the perceiving subject is the basis of all natural science. Since, however, sense perception only gives information of this external world or of "physical reality" indirectly, we can only grasp the latter by speculative means. It follows from this that our notions of physical reality can never be final. We must always be ready to change these notions—that is to say, the axiomatic sub-structure of physics—in order to do justice to perceived facts in the most logically perfect way. Actually a glance at the development of physics shows that it has undergone far-reaching changes in the course of time.

The greatest change in the axiomatic sub-structure of physics—in other words, of our conception of the structure of reality—since Newton laid the foundation of theoretical physics was brought about by Faraday's and Clerk Maxwell's work on electromagnetic phenomena. We will try in what follows to make this clearer, keeping both earlier and later developments in sight.

According to Newton's system, physical reality is

characterized by the concepts of time, space, material point, and force (reciprocal action of material points). Physical events, in Newton's view, are to be regarded as the motions, governed by fixed laws, of material points in space. The material point is our only mode of representing reality when dealing with changes taking place in it, the solitary representative of the real, in so far as the real is capable of change. Perceptible bodies are obviously responsible for the concept of the material point; people conceived it as an analogue of mobile bodies, stripping these of the characteristics of extension, form, orientation in space, and all "inward" qualities, leaving only inertia and translation and adding the concept of force. The material bodies, which had led psychologically to our formation of the concept of the "material point," had now themselves to be regarded as systems of material points. It should be noted that this theoretical scheme is in essence an atomistic and mechanistic one. All happenings were to be interpreted purely mechanically—that is to say, simply as motions of material points according to Newton's law of motion.

The most unsatisfactory side of this system (apart from the difficulties involved in the concept of "absolute space" which have been raised once more quite recently) lay in its description of light, which Newton also conceived, in accordance with his system, as composed of material points. Even at that time the question, What in that case becomes of the material points of which light is composed, when the light is absorbed? was already a burning one. Moreover, it is unsatisfactory in any case to introduce into the discussion material points of quite a different sort, which had to be postulated for the purpose of representing ponderable matter and light respectively. Later on electrical corpuscles were added to these, making a third kind, again with completely different characteristics. It was, further, a fundamental weakness that the forces of reciprocal action, by which events are determined, had to be assumed hypothetically in a perfectly arbitrary way. Yet this conception of the real accomplished much: how came it that people felt themselves impelled to forsake it?

In order to put his system into mathematical form at all, Newton had to devise the concept of differential quotients and propound the laws of motion in the form of total differential equations—perhaps the greatest advance in thought that a single individual was ever privileged to make. Partial differential equations were not necessary for this purpose, nor did Newton make any systematic use of them; but they were necessary for the formulation of the mechanics of deformable bodies; this is connected with the fact that in these problems the question of *how* bodies are supposed to be constructed out of material points was of no importance to begin with.

Thus the partial differential equation entered theoretical physics as a handmaid, but has gradually become mistress. This began in the nineteenth century when the wave-theory of light established itself under the pressure of observed fact. Light in empty space was explained as a matter of vibrations of the ether, and it seemed idle at that stage, of course, to look upon the latter as a conglomeration of material points. Here for the first time the partial differential equation appeared as the natural expression of the primary realities of physics. In a particular department of theoretical physics the continuous field thus appeared side by side with the material point as the representative of physical reality. This dualism remains even today, disturbing as it must be to every orderly mind.

If the idea of physical reality had ceased to be purely atomic, it still remained for the time being purely *mechanistic*; people still tried to explain all events as the motion of inert masses; indeed no other way of looking at things seemed conceivable. Then came the great change, which will be associated for all time with the names of Faraday, Clerk Maxwell, and Hertz. The lion's share in this revolution fell to Clerk Maxwell. He showed that the whole of what was then known about light and electro-magnetic phenomena was expressed in his well known double system of differential equations, in which the electric and the magnetic fields appear as the dependent variables. Maxwell did, indeed, try to explain, or justify, these equations by intellectual constructions.

But he made use of several such constructions at the same time and took none of them really seriously, so that the equations alone appeared as the essential thing and the strength of the fields as the ultimate entities, not to be reduced to anything else. By the turn of the century the conception of the electromagnetic field as an ultimate entity had been generally accepted and serious thinkers had abandoned the belief in the justification, or the possibility, of a mechanical explanation of Clerk Maxwell's equations. Before long they were, on the contrary, actually trying to explain material points and their inertia on field theory lines with the help of Maxwell's theory, an attempt which did not, however, meet with complete success.

Neglecting the important *individual* results which Clerk Maxwell's life-work produced in important departments of physics, and concentrating on the changes wrought by him in our conception of the nature of physical reality, we may say this:—Before Clerk Maxwell people conceived of physical reality—in so far as it is supposed to represent events in nature—as material points, whose changes consist exclusively of motions, which are subject to partial differential equations. After Maxwell they conceived physical reality as represented by continuous fields, not mechanically explicable, which are subject to partial differential equations. This change in the conception of reality is the most profound and fruitful one that has come to physics since Newton; but it has at the same time to be admitted that the program has by no means been completely carried out yet. The successful systems of physics which have been evolved since rather represent compromises between these two schemes, which for that very reason bear a provisional, logically incomplete character, although they may have achieved great advances in certain particulars.

The first of these that calls for mention is Lorentz's

theory of electrons, in which the field and the electrical corpuscles appear side by side as elements of equal value for the comprehension of reality. Next come the special and general theories of relativity, which, though based entirely on ideas connected with the field-theory, have so far been unable to avoid the independent introduction of material points and total differential equations.

The last and most successful creation of theoretical physics, namely quantum-mechanics, differs fundamentally from both the schemes which we will for the sake of brevity call the Newtonian and the Maxwellian. For the quantities which figure in its laws make no claim to describe physical reality itself, but only the *probabilities* of the occurrence of a physical reality that we have in view. Dirac, to whom, in my opinion, we owe the most logically complete exposition of this theory, rightly points out that it would probably be difficult, for example, to give a theoretical description of a photon such as would give enough information to enable one to decide whether it will pass a polarizer placed (obliquely) in its way or not.

I am still inclined to the view that physicists will not in the long run content themselves with that sort of indirect description of the real, even if the theory can eventually be adapted to the postulate of general relativity in a satisfactory manner. We shall then, I feel sure, have to return to the attempt to carry out the program which may properly be described as the Maxwellian—namely, the description of physical reality in terms of fields which satisfy partial differential equations without singularities.

NIELS BOHR

When a later generation comes to write the history of the progress made in physics in our time, it will have to connect one of the most important advances ever made in our knowledge of the nature of the atom with the name of Niels Bohr. It was already known that classical mechanics break down in relation to the ultimate constituents of matter, also that atoms consist of positively charged nuclei which are surrounded by a layer of atoms of relatively rather loose texture. But the structure of the spectra, which was to a large extent known empirically, was so profoundly different from what was to be expected on our older theories that nobody could find a convincing theoretical interpretation of the observed uniformities. Thereupon Bohr in the year 1913 devised an interpretation of the simplest spectra on quantum-theory lines, for which he in a short time produced such a mass of quantitative confirmation that the boldly selected hypothetical basis of his speculations soon became a mainstay for the physics of the atom. Although less than ten years have passed since Bohr's first discovery, the system conceived in its main features and largely worked out by him already dominates both physics and chemistry so completely that all earlier systems seem to the expert to date from a long vanished age. The theory of the Röntgen spectra, of the visible spectra, and the periodic system of the elements is primarily based on the ideas of Bohr. What is so marvelously attractive about Bohr as a scientific thinker is his rare blend of boldness and caution; seldom has anyone possessed such an intuitive grasp of hidden things combined with such a strong critical sense. With all his knowledge of the details, his eye is immovably fixed on the underlying principle. He is unquestionably one of the greatest discoverers of our age in the scientific field.

ON THE THEORY OF RELATIVITY

It is a particular pleasure to me to have the privilege of speaking in the capital of the country, from which the most important fundamental notions of theoretical physics have issued. I am thinking of the theory of mass motion and gravitation which Newton gave us and the concept of the electro-magnetic field, by means of which Faraday and Clerk Maxwell put physics on a new basis. The theory of relativity may indeed be said to have put a sort of finishing touch to the mighty intellectual edifice of Maxwell and Lorentz, inasmuch as it seeks to extend field physics to all phenomena, gravitation included.

Turning to the theory of relativity itself, I am anxious to draw attention to the fact that this theory is not speculative in origin; it owes its invention entirely to the desire to make physical theory fit observed fact as well as possible. We have here no revolutionary act but the natural continuation of a line that can be traced through centuries. The abandonment of a certain concept connected with space, time and motion hitherto treated as fundamentals must not be regarded as arbitrary, but only as conditioned by observed facts.

The law of the constant velocity of light in empty

space, which has been confirmed by the development of electro-dynamics and optics, and the equal legitimacy of all inertial systems (special principle of relativity), which was proved in a particularly incisive manner by Michelson's famous experiment, between them made it necessary, to begin with, that the concept of time should be made relative, each inertial system being given its own special time. As this notion was developed it became clear that the connection between immediate experience on one side and co-ordinates and time on the other had hitherto not been thought out with sufficient precision. It is in general one of the essential features of the theory of relativity that it is at pains to work out the relations between general concepts and empirical facts more precisely. The fundamental principle here is that the justification for a physical concept lies exclusively in its clear and unambiguous relation to facts that can be experienced. According to the special theory of relativity, spatial co-ordinates and time still have an absolute character in so far as they are directly measurable by stationary clocks and bodies. But they are relative in so far as they depend on the state of motion of the selected inertial system. According to the special theory of relativity the four-dimensional continuum formed by the union of space and time retains the absolute character which, according to the earlier theory, belonged to both space and time separately (Minkowski). The influence of motion (relative to the co-ordinate system) on the form of bodies and on the motion of clocks, also the equivalence of energy and inert mass, follow from the interpretation of co-ordinates and time as products of measurement.

The general theory of relativity owes its existence in the first place to the empirical fact of the numerical equality of the inertial and gravitational mass of bodies, for which fundamental fact classical mechanics provided no interpretation. Such an interpretation is arrived at by an extension of the principle of relativity to co-ordinate systems accelerated relatively to one another. The introduction of co-ordinate systems accelerated relatively to inertial systems involves the appearance of gravitational fields relative to the latter. As a result of this, the general theory of relativity, which is based on the equality of inertia and weight, provides a theory of the gravitational field.

The introduction of co-ordinate systems accelerated relatively to each other as equally legitimate systems, such as they appear conditioned by the identity of inertia and weight, leads, in conjunction with the results of the special theory of relativity, to the conclusion that the laws governing the occupation of space by solid bodies, when gravitational fields are present, do not correspond to the laws of Euclidean geometry. An analogous result follows from the motion of clocks. This brings us to the necessity for yet another generalization of the theory of space and time, because the direct interpretation of spatial and temporal co-ordinates by means of measurements obtainable with measuring rods and clocks now breaks down. That generalization of metric, which had already been accomplished in the sphere of pure mathematics through the researches of Gauss and Riemann, is essentially based on the fact that the metric of the special theory of relativity can still claim validity for small areas in the general case as well.

The process of development here sketched strips the space-time co-ordinates of all independent reality. The metrically real is now only given through the combination of the space-time co-ordinates with the mathematical quantities which describe the gravitational field.

There is yet another factor underlying the evolution of the general theory of relativity. As Ernst Mach insistently pointed out, the Newtonian theory is unsatisfactory in the following respect:—If one considers motion from the purely descriptive, not from the causal, point of view, it only exists as relative motion of things with respect to one another. But the acceleration which figures in Newton's equations of motion is unintelligible if one starts with the concept of relative motion. It compelled Newton to invent a physical space in relation to which acceleration was supposed to exist. This introduction *ad hoc* of the concept of absolute space, while logically unexceptionable, nevertheless seems unsatisfactory. Hence the attempt to alter the mechanical equations in such a way that the inertia of bodies is traced back to relative motion on their part not as against absolute space but as against the totality of other ponderable bodies. In the state of knowledge then existing his attempt was bound to fail.

The posing of the problem seems, however, entirely reasonable. This line of argument imposes itself with considerably enhanced force in relation to the general theory of relativity, since, according to that theory, the physical properties of space are affected by ponderable matter. In my opinion the general theory of relativity can only solve this problem satisfactorily if it regards the world as spatially self-enclosed. The mathematical results of the theory force one to this view, if one believes that the mean density of ponderable matter in the world possesses some ultimate value, however small.

(An address in London)

WHAT IS THE THEORY OF RELATIVITY?

I GLADLY accede to the request of your colleague to write something for *The Times* on relativity. After the lamentable breakdown of the old active intercourse between men of learning, I welcome this opportunity of expressing my feelings of joy and gratitude towards the astronomers and physicists of England. It is thoroughly in keeping with the great and proud traditions of scientific work in your country that eminent scientists should have spent much time and trouble, and your scientific institutions have spared no expense, to test the implications of a theory which was perfected and published during the War in the land of your enemies. Even though the investigation of the influence of the gravitational field of the sun on light rays is a purely objective matter, I cannot forbear to express my personal thanks to my English colleagues for their work; for without it I could hardly have lived to see the most important implication of my theory tested.

We can distinguish various kinds of theories in physics. Most of them are constructive. They attempt to build up a picture of the more complex phenomena out of the materials of a relatively simple formal scheme from which they start out. Thus the kinetic theory of gases seeks to reduce mechanical, thermal and diffusional processes to movements of molecules —i.e., to build them up out of the hypothesis of molecular motion. When we say that we have succeeded in understanding a group of natural processes, we invariably mean that a constructive theory has been found which covers the processes in question.

Along with this most important class of theories there exists a second, which I will call "principle-theories." These employ the analytic, not the synthetic, method. The elements which form their basis and starting-point are not hypothetically constructed but empirically discovered ones, general characteristics of natural processes, principles that give rise to mathematically formulated criteria which the separate processes or the theoretical representations of them have to satisfy. Thus the science of thermodynamics seeks by analytical means to deduce necessary connections, which separate events have to satisfy, from the universally experienced fact that perpetual motion is impossible.

The advantages of the constructive theory are completeness, adaptability and clearness, those of the principle theory are logical perfection and security of the foundations.

The theory of relativity belongs to the latter class. In order to grasp its nature, one needs first of all to become acquainted with the principles on which it is based. Before I go into these, however, I must observe that the theory of relativity resembles a building consisting of two separate stories, the special theory and the general theory. The special theory, on which the general theory rests, applies to all physical phenomena with the exception of gravitation; the general theory provides the law of gravitation and its relations to the other forces of nature.

It has, of course, been known since the days of the ancient Greeks that in order to describe the movement of a body, a second body is needed to which the movement of the first is referred. The movement of a vehicle is considered in reference to the earth's surface, that of a planet to the totality of the visible fixed stars. In physics the body to which events are spatially referred is called the co-ordinate system. The laws of the mechanics of Galileo and Newton, for instance, can only be formulated with the aid of a co-ordinate system.

The state of motion of the co-ordinate system may not, however, be arbitrarily chosen, if the laws of mechanics are to be valid (it must be free from rotation and acceleration). A co-ordinate system which is admitted in mechanics is called an "inertial system." The state of motion of an inertial system is according to mechanics not one that is determined uniquely by nature. On the contrary, the following definition holds good:—a co-ordinate system that is moved uniformly and in a straight line relatively to an inertial system is likewise an inertial system. By the "special principle of relativity" is meant the generalization of this definition to include any natural event whatever: thus, every universal law of nature which is valid in relation to a co-ordinate system C, must also be valid, as it stands, in relation to a co-ordinate system C′, which is in uniform translatory motion relatively to C.

The second principle, on which the special theory of relativity rests, is the "principle of the constant velocity of light in vacuo." This principle asserts that light in vacuo always has a definite velocity of propagation (independent of the state of motion of the observer or of the source of the light). The confidence which physicists place in this principle springs from the successes achieved by the electro-dynamics of Clerk Maxwell and Lorentz.

Both the above-mentioned principles are powerfully supported by experience, but appear not to be logically reconcilable. The special theory of relativity finally succeeded in reconciling them logically by a modification of kinematics—i.e., of the doctrine of the laws relating to space and time (from the point of view of physics). It became clear that to speak of the simultaneity of two events had no meaning except in relation to a given co-ordinate system, and that the shape of measuring devices and the speed at which clocks move depend on their state of motion with respect to the co-ordinate system.

But the old physics, including the laws of motion of Galileo and Newton, did not fit in with the suggested relativist kinematics. From the latter, general mathematical conditions issued, to which natural laws had to conform, if the above-mentioned two principles were really to apply. To these, physics had to be adapted. In particular, scientists arrived at a new law of motion for (rapidly moving) mass points, which was admirably confirmed in the case of electrically charged particles. The most important upshot of the special theory of relativity concerned the inert mass of corporeal systems. It turned out that the inertia of a system necessarily depends on its energy-content, and this led straight to the notion that inert mass is simply latent energy. The principle of the conservation of mass lost its independence and became fused with that of the conservation of energy.

The special theory of relativity, which was simply a systematic development of the electro-dynamics of Clerk Maxwell and Lorentz, pointed beyond itself, however. Should the independence of physical laws of the state of motion of the co-ordinate system be restricted to the uniform translatory motion of co-ordinate systems in respect to each other? What has nature to do with our co-ordinate systems and their state of motion? If it is necessary for the purpose of describing nature, to make use of a co-ordinate system arbitrarily introduced by us, then the choice of its state of motion ought to be subject to no restriction; the laws ought to be entirely independent of this choice (general principle of relativity).

The establishment of this general principle of relativity is made easier by a fact of experience that has long been known, namely that the weight and the inertia of a body are controlled by the same constant. (Equality of inertial and gravitational mass.) Imagine a co-ordinate system which is rotating uniformly with respect to an inertial system in the Newtonian manner. The centrifugal forces which manifest themselves in relation to this system must, according to Newton's teaching, be regarded as effects of inertia. But these centrifugal forces are, exactly like the forces of gravity, proportional to the masses of the bodies. Ought it not to be possible in this case to regard the co-ordinate system as stationary and the centrifugal forces as gravitational forces? This seems the obvious view, but classical mechanics forbid it.

This hasty consideration suggests that a general theory of relativity must supply the laws of gravitation, and the consistent following up of the idea has justified our hopes.

But the path was thornier than one might suppose, because it demanded the abandonment of Euclidean geometry. This is to say, the laws according to which fixed bodies may be arranged in space, do not completely accord with the spatial laws attributed to bodies by Euclidean geometry. This is what we mean when we talk of the "curvature of space." The fundamental concepts of the "straight line," the "plane," etc., thereby lose their precise significance in physics.

In the general theory of relativity the doctrine of space and time, or kinematics, no longer figures as a fundamental independent of the rest of physics. The geometrical behavior of bodies and the motion of clocks rather depend on gravitational fields, which in their turn are produced by matter.

The new theory of gravitation diverges considerably, as regards principles, from Newton's theory. But its practical results agree so nearly with those of Newton's theory that it is difficult to find criteria for distinguishing them which are accessible to experience. Such have been discovered so far:—

(1) In the revolution of the ellipses of the planetary orbits round the sun (confirmed in the case of Mercury).
(2) In the curving of light rays by the action of gravitational fields (confirmed by the English photographs of eclipses).
(3) In a displacement of the spectral lines towards the red end of the spectrum in the case of light transmitted to us from stars of considerable magnitude (unconfirmed so far).[1]

The chief attraction of the theory lies in its logical completeness. If a single one of the conclusions drawn from it proves wrong, it must be given up; to modify it without destroying the whole structure seems to be impossible.

Let no one suppose, however, that the mighty work of Newton can really be superseded by this or any other theory. His great and lucid ideas will retain their unique significance for all time as the foundation of our whole modern conceptual structure in the sphere of natural philosophy.

[1] Editor's Note: This criterion has also been confirmed in the meantime.

NOTE: Some of the statements in your paper concerning my life and person owe their origin to the lively imagination of the writer. Here is yet another application of the principle of relativity for the delectation of the reader:—Today I am described in Germany as a "German savant," and in England as a "Swiss Jew." Should it ever be my fate to be represented as a *bête noire,* I should, on the contrary, become a "Swiss Jew" for the Germans and a "German savant" for the English.

THE PROBLEM OF SPACE, ETHER, AND THE FIELD IN PHYSICS

SCIENTIFIC thought is a development of pre-scientific thought. As the concept of space was already fundamental in the latter, we must begin with the concept of space in pre-scientific thought. There are two ways of regarding concepts, both of which are necessary to understanding. The first is that of logical analysis. It answers the question, How do concepts and judgments depend on each other? In answering it we are on comparatively safe ground. It is the security by which we are so much impressed in mathematics. But this security is purchased at the price of emptiness of content. Concepts can only acquire content when they are connected, however indirectly, with sensible experience. But no logical investigation can reveal this connection; it can only be experienced. And yet it is this connection that determines the cognitive value of systems of concepts.

Take an example. Suppose an archaeologist belonging to a later culture finds a text-book of Euclidean geometry without diagrams. He will discover how the words "point," "straight-line," "plane" are used in the propositions. He will also see how the latter are deduced from each other. He will even be able to frame new propositions according to the known rules. But the framing of these propositions will remain an empty word-game for him, as long as "point," "straight-line," "plane," etc., convey nothing to him. Only when they do convey something will geometry possess any real content for him. The same will be true of analytical mechanics, and indeed of any exposition of the logically deductive sciences.

What does this talk of "straight-line," "point," "intersection," etc., conveying something to one, mean? It means that one can point to the parts of sensible experience to which those words refer. This extra-logical problem is the essential problem, which the archaeologist will only be able to solve intuitively, by examining his experience and seeing if he can discover anything which corresponds to those primary terms of the theory and the axioms laid down for them. Only in this sense can the question of the nature of a conceptually presented entity be reasonably raised.

With our pre-scientific concepts we are very much in the position of our archaeologist in regard to the ontological problem. We have, so to speak, forgotten what features in the world of experience caused us to frame those concepts, and we have great difficulty in representing the world of experience to ourselves without the spectacles of the old-established conceptual interpretation. There is the further difficulty that our language is compelled to work with words which are inseparably connected with those primitive concepts. These are the obstacles which confront us when we try to describe the essential nature of the pre-scientific concept of space.

One remark about concepts in general, before we turn to the problem of space: concepts have reference to sensible experience, but they are never, in a logical sense, deducible from them. For this reason I have never been able to understand the quest of the *a priori* in the Kantian sense. In any ontological question, the only possible procedure is to seek out these characteristics in the complex of sense experiences to which the concepts refer.

Now as regards the concept of space: this seems to presuppose the concept of the solid object. The nature of the complexes and sense-impressions which are probably responsible for that concept has often been described. The correspondence between certain visual and tactile impressions, the fact that they can be continuously followed out through time, and that the impressions can be repeated at any movement (taste, sight), are some of those characteristics. Once the concept of the solid object is formed in connection with the experiences just mentioned—which concept by no means presupposes that of space or spatial relation—the desire to get an intellectual grasp of the relations of such solid bodies is bound to give rise to concepts which correspond to their spatial relations. Two solid objects may touch one another or be distant from one another. In the latter case, a third body can be inserted between them without altering them in any way, in the former not. These spatial relations are obviously real in the same sense as the bodies themselves. If two bodies are of equal value for the filling of *one* such interval, they will also prove of equal value for the filling of other intervals. The interval is thus shown to be independent of the selection of any special body to fill it; the same is universally true of spatial relations. It is plain that this independence, which is a principle condition of the usefulness of framing purely geometrical concepts, is not necessary *a priori*. In my opinion, this concept of the interval, detached as it is from the selection of any special body to occupy it, is the starting point of the whole concept of space.

Considered, then, from the point of view of sense experience, the development of the concept of space seems, after these brief indications, to conform to the following schema—solid body; spatial relations of solid bodies; interval; space. Looked at in this way, space appears as something real in the same sense as solid bodies.

It is clear that the concept of space as a real thing already existed in the extra-scientific conceptual world. Euclid's mathematics, however, knew nothing of this concept as such; they confined themselves to the concepts of the object, and the spatial relations between objects. The point, the plane, the straight line, length, are solid objects idealized. All spatial relations are reduced to those of contact (the intersection of straight lines and planes, points lying on straight lines, etc.). Space as a continuum does not figure in the conceptual system at all. This concept was first introduced by Descartes, when he described the point-in-space by its co-ordinates. Here for the first time geometrical figures appear, up to a point, as parts of infinite space, which is conceived as a three-dimensional continuum.

The great superiority of the Cartesian treatment of space is by no means confined to the fact that it applies analysis to the purposes of geometry. The main point seems rather to be this:—The geometry of the Greeks prefers certain figures (the straight line, the plane) in geometrical descriptions; other figures (e.g., the ellipse) are only accessible to it because it constructs or defines them with the help of the point, the straight line and the plane. In the Cartesian treatment on the other hand, all surfaces are, in principle, equally represented, without any arbitrary preference for linear figures in the construction of geometry.

In so far as geometry is conceived as the science of laws governing the mutual relations of practically rigid bodies in space, it is to be regarded as the oldest branch of physics. This science was able, as I have already observed, to get along without the concept of space as such, the ideal corporeal forms—point, straight line, plane, length—being sufficient for its needs. On the other hand, space as a whole, as conceived by Descartes, was absolutely necessary to Newtonian physics. For dynamics cannot manage with the concepts of the mass point and the (temporally variable) distance between mass points alone. In Newton's equations of motion the concept of acceleration plays a fundamental part, which cannot be defined by the temporally variable intervals between points alone. Newton's acceleration is only thinkable or definable in relation to space as a whole. Thus to the geometrical reality of the concept of space a new inertia-determining function of space was added. When Newton described space as absolute, he no doubt meant this real significance of space, which made it necessary for him to attribute to it a quite definite state of motion, which yet did not appear to be fully determined by the phenomena of mechanics. This space was conceived as absolute in another sense also; its inertia-determining effect was conceived as autonomous, i.e., not to be influenced by any physical circumstance whatever; it affected masses, but nothing affected it.

And yet in the minds of physicists space remained until the most recent time simply the passive container of all events, playing no part in physical happenings itself. Thought only began to take a new turn with the wave theory of light and the theory of the electromagnetic field of Faraday and Clerk Maxwell. It became clear that there existed in free space conditions which propagated themselves in waves, as well as localized fields which were able to exert force on electrical masses or magnetic poles brought to the spot. Since it would have seemed utterly absurd to the physicists of the nineteenth century to attribute physical functions or states to space itself, they invented a medium pervading the whole of space, on the model of ponderable matter—the ether, which was supposed to act as a vehicle for electro-magnetic phenomena, and hence for those of light also. The states of this medium, imagined as constituting the electro-magnetic fields, were at first thought of mechanically, on the model of the elastic deformations of rigid bodies. But this mechanical theory of the ether was never quite successful and so the idea of a closer explanation of the nature of the etheric fields was given up. The ether thus became a kind of matter whose only function was to act as a substratum for electrical fields which were by their very nature not further analyzable. The picture was, then, as follows:—Space is filled by the ether, in which the material corpuscles or atoms of ponderable matter swim; the atomic structure of the latter had been securely established by the turn of the century.

Since the reciprocal action of bodies was supposed to be accomplished through fields, there had also to be a gravitational field in the ether, whose field-law had, however, assumed no clear form at that time. The ether was only accepted as the seat of all operations of force which make themselves effective across space. Since it had been realized that electrical masses in motion produce a magnetic field, whose energy acted as a model for inertia, inertia also appeared as a field-action localized in the ether.

The mechanical properties of the ether were at first a mystery. Then came H. A. Lorentz's great discovery. All the phenomena of electro-magnetism then known could be explained on the basis of two assumptions: that the ether is firmly fixed in space —that is to say, unable to move at all, and that electricity is firmly lodged in the mobile elementary particles. Today his discovery may be expressed as follows:—Physical space and the ether are only different terms for the same thing; fields are physical conditions of space. For if no particular state of motion belongs to the ether, there does not seem to be any ground for introducing it as an entity of a special sort alongside of space. But the physicists were still far removed from such a way of thinking; space was still, for them, a rigid, homogeneous something, susceptible of no change or conditions. Only the genius of Riemann, solitary and uncomprehended, had already won its way by the middle of last century to a new conception of space, in which space was deprived of its rigidity, and in which its power to take part in physical events was recognized as possible. This intellectual achievement commands our admiration all the more for having preceded Faraday's and Clerk Maxwell's field theory of electricity. Then came the special theory of relativity with its recognition of the physical equivalence of all inertial systems. The inseparableness of time and space emerged in connection with electrodynamics, or the law of the propagation of light. Hitherto it had been silently assumed that the four-dimensional continuum of events could be split up into time and space in an objective manner—i.e., that an absolute significance attached to the "now" in the world of events. With the discovery of the relativity of simultaneity, space and time were merged in a single continuum in the same way as the three-dimensions of space had been before. Physical space was thus increased to a four-dimensional space which also included the dimension of time. The four-dimensional space of the special theory of relativity is just as rigid and absolute as Newton's space.

The theory of relativity is a fine example of the fundamental character of the modern development of theoretical science. The hypotheses with which it starts become steadily more abstract and remote from experience. On the other hand it gets nearer to the grand aim of all science, which is to cover the greatest possible number of empirical facts by logical deduction from the smallest possible number of hypotheses or axioms. Meanwhile the train of thought leading from the axioms to the empirical facts or verifiable consequences gets steadily longer and more subtle. The theoretical scientist is compelled in an increasing degree to be guided by purely mathematical, formal considerations in his search for a theory, because the physical experience of the experimenter cannot lift him into the regions of highest abstraction. The predominantly inductive methods appropriate to the youth of science are giving place to tentative deduction. Such a theoretical structure needs to be very thoroughly elaborated before it can lead to conclusions which can be compared with experience. Here too the observed fact is undoubtedly the supreme arbiter; but it cannot pronounce sentence until the wide chasm separating the axioms from their verifiable consequences has been bridged by much intense, hard thinking. The theorist has to set about this Herculean task in the clear consciousness that his efforts may only be destined to deal the death blow to his theory. The theorist who undertakes such a labor should not be carped at as "fanciful"; on the contrary, he should be encouraged to give free reign to his fancy, for there is no other way to the goal. His is no idle daydreaming, but a search for the logically simplest possibilities and their consequences. This plea was needed in order to make the hearer or reader more ready to follow the ensuing train of ideas with attention; it is the line of thought which has led from the special to the general theory of relativity and thence to its latest offshoot, the unitary field theory. In this exposition the use of mathematical symbols cannot be avoided.

We start with the special theory of relativity. This theory is still based directly on an empirical law, that of the constant velocity of light. Let P be a point in empty space, P′ one separated from it by a length $d\sigma$ and infinitely near to it. Let a flash of light be emitted from P at a time t and reach P′ at a time $t+dt$. Then

$$d\sigma^2 = C^2 dt^2$$

If dx_1, dx_2, dx_3 are the orthogonal projections of $d\sigma$, and the imaginary time co-ordinate $\sqrt{-1}ct = x_4$ is introduced, then the above-mentioned law of the constancy of the propagation of light takes the form

$$ds^2 = dx_1{}^2 + dx_2{}^2 + dx_3{}^2 + dx_4{}^2 = 0$$

Since this formula expresses a real situation, we may attribute a real meaning to the quantity ds, even supposing the neighboring points of the four-dimensional continuum are selected in such a way that the ds belonging to them does not disappear. This is more or less expressed by saying that the four-dimensional space (with imaginary time-co-ordinates) of the special theory of relativity possesses a Euclidean metric.

The fact that such a metric is called Euclidean is connected with the following. The position of such a metric in a three-dimensional continuum is fully equivalent to the positions of the axioms of Euclidean geometry. The defining equation of the metric is thus nothing but the Pythagorean theorem applied to the differentials and the co-ordinates.

Such alteration of the co-ordinates (by transformation) is permitted in the special theory of relativity, since in the new co-ordinates too the magnitude ds^2 (fundamental invariant) is expressed in the new differentials of the co-ordinates by the sum of the squares. Such transformations are called Lorentz transformations.

The leuristic method of the special theory of relativity is characterized by the following principle:—Only those equations are admissible as an expression of natural laws which do not change their form when the co-ordinates are changed by means of a Lorentz transformation (co-variance of equations in relation to Lorentz transformations).

This method led to the discovery of the necessary connection between impulse and energy, the strength of an electric and a magnetic field, electrostatic and electro-dynamic forces, inert mass and energy; and the number of independent concepts and fundamental equations was thereby reduced.

This method pointed beyond itself. Is it true that the equations which express natural laws are covariant in relation to Lorentz transformations only and not in relation to other transformations? Well, formulated in that way the question really means nothing, since every system of equations can be expressed in general co-ordinates. We must ask, Are not the laws of nature so constituted that they receive no real simplification through the choice of any one *particular* set of co-ordinates?

We will only mention in passing that our empirical principle of the equality of inert and heavy masses prompts us to answer this question in the affirmative. If we elevate the equivalence of all co-ordinate systems for the formulation of natural laws into a principle, we arrive at the general theory of relativity, provided we stick to the law of the constant velocity of light or to the hypothesis of the objective significance of the Euclidean metric at least for infinitely small portions of four-dimensional space.

This means that for finite regions of space the existence (significant for physics) of a general Riemannian metric is presupposed according to the formula

$$ds^2 = \sum_{\mu\nu} g_{\mu\nu} \, dx^\mu \, dx^\nu,$$

whereby the summation is to be extended to all index combinations from 11 to 44.

The structure of such a space differs absolutely radically in *one* respect from that of a Euclidean space. The coefficients $g_{\mu\nu}$ are for the time being any functions whatever of the co-ordinates x_1 to x_4, and the structure of the space is not really determined until these functions $g_{\mu\nu}$ are really known. It is only determined more closely by specifying laws which the metrical field of the $g_{\mu\nu}$ satisfy. On physical grounds this gave rise to the conviction that the metrical field was at the same time the gravitational field.

Since the gravitational field is determined by the configuration of masses and changes with it, the geometric structure of this space is also dependent on physical factors. Thus according to this theory space is—exactly as Riemann guessed—no longer absolute; its structure depends on physical influences. Physical geometry is no longer an isolated self-contained science like the geometry of Euclid.

The problem of gravitation was thus reduced to a mathematical problem: it was required to find the simplest fundamental equations which are co-variant in relation to any transformation of co-ordinates whatever.

I will not speak here of the way this theory has been confirmed by experience, but explain at once why Theory could not rest permanently satisfied with this success. Gravitation had indeed been traced to the structure of space, but besides the gravitational field there is also the electro-magnetic field. This had, to begin with, to be introduced into the theory as an entity independent of gravitation. Additional terms which took account of the existence of the electromagnetic field had to be included in the fundamental equations for the field. But the idea that there were two structures of space independent of each other, the metric-gravitational and the electro-magnetic, was intolerable to the theoretical spirit. We are forced to the belief that both sorts of field must correspond to verified structure of space.

The "unitary field-theory," which represents itself as a mathematically independent extension of the general theory of relativity, attempts to fulfil this last postulate of the field theory. The formal problem should be put as follows:—Is there a theory of the continuum in which a new structural element appears side by side with the metric such that it forms a single whole together with the metric? If so, what are the simplest field laws to which such a continuum can be made subject? And finally, are these field-laws well fitted to represent the properties of the gravitational field and the electromagnetic field? Then there is the further question whether the corpuscles (electrons and protons) can be regarded as positions of particularly dense fields, whose movements are determined by the field equations. At present there is only one way of answering the first three questions. The space structure on which it is based may be described as follows, and the description applies equally to a space of any number of dimensions.

Space has a Riemannian metric. This means that the Euclidean geometry holds good in the infinitesimal neighborhood of every point P. Thus for the neighborhood of every point P there is a local Cartesian system of co-ordinates, in reference to which the metric is calculated according to the Pythagorean theorem. If we now imagine the length 1 cut off from the positive axes of these local systems, we get the orthogonal "local n-leg." Such a local n-leg is to be found in every other point P′ of space also. Thus, if a linear element (PG or P′G′) starting from the points P or P′, is given, then the magnitude of this linear element can be calculated by the aid of the relevant local n-leg from its local co-ordinates by means of Pythagoras's theorem. There is therefore a definite meaning in speaking of the numerical equality of the linear elements PG and P′G′.

It is essential to observe now that the local orthogonal n-legs are not completely determined by the metric. For we can still select the orientation of the n-legs perfectly freely without causing any alteration in the result of calculating the size of the linear elements according to Pythagoras's theorem. A corollary of this is that in a space whose structure consists exclusively of a Riemannian metric, two linear elements PG and P′G′, can be compared with regard to their magnitude but not their direction; in particular, there is no sort of point in saying that the two linear elements are parallel to one another. In this respect, therefore, the purely metrical (Riemannian) space is less rich in structure than the Euclidean.

Since we are looking for a space which exceeds Riemannian space in wealth of structure, the obvious thing is to enrich Riemannian space by adding the relation of direction or parallelism. Therefore for every direction through P let there be a definite direction through P′, and let this mutual relation be a determinate one. We call the directions thus related to each other "parallel." Let this parallel relation further fulfil the condition of angular uniformity: If PG and PK are two directions in P, P′G′ and P′K′ the corresponding parallel directions through P′, then the angles KPG and K′P′G′ (measurable on Euclidean lines in the local system) should be equal.

The basic space-structure is thereby completely defined. It is most easily described mathematically as follows:—In the definite point P we suppose an orthogonal n-leg with definite, freely chosen orientation. In every other point P′ of space we so orient its local n-leg that its axes are parallel to the corresponding axes at the point P. Given the above structure of space and free choice in the orientation of the n-leg at one point P, all n-legs are thereby completely defined. In the space P let us now imagine any Gaussian system of co-ordinates and that in every point the axes of the n-leg there are projected on to it. This system of n^2 components completely describes the structure of space.

This spatial structure stands, in a sense, midway between the Riemannian and the Euclidean. In contrast to the former, it has room for the straight-line, that is to say a line all of whose elements are parallel to each other in pairs. The geometry here described differs from the Euclidean in the non-existence of the parallelogram. If at the ends P and G of a length

PG two equal and parallel lengths PP′ and GG′ are marked off, P′G′ is in general neither equal nor parallel to PG.

The mathematical problem now solved so far is this:—What are the simplest conditions to which a space-structure of the kind described can be subjected? The chief question which still remains to be investigated is this:—To what extent can physical fields and primary entities be represented by solutions, free from singularities, of the equations which answer the former question?

NOTES ON THE ORIGIN OF THE GENERAL THEORY OF RELATIVITY

I GLADLY accede to the request that I should say something about the history of my own scientific work. Not that I have an exaggerated notion of the importance of my own efforts, but to write the history of other men's work demands a degree of absorption in other people's ideas which is much more in the line of the trained historian; to throw light on one's own earlier thinking appears incomparably easier. Here one has an immense advantage over everybody else, and one ought not to leave the opportunity unused out of modesty.

When, by the special theory of relativity I had arrived at the equivalence of all so-called inertial systems for the formulation of natural laws (1905), the question whether there was not a further equivalence of co-ordinate systems followed naturally, to say the least of it. To put it in another way, if only a relative meaning can be attached to the concept of velocity, ought we nevertheless to persevere in treating acceleration as an absolute concept?

From the purely kinematic point of view there was no doubt about the relativity of all motions whatever; but physically speaking, the inertial system seemed to occupy a privileged position, which made the use of co-ordinate systems moving in other ways appear artificial.

I was of course acquainted with Mach's view, according to which it appeared conceivable that what inertial resistance counteracts is not acceleration as such but acceleration with respect to the masses of the other bodies existing in the world. There was something fascinating about this idea to me, but it provided no workable basis for a new theory.

I first came a step nearer to the solution of the problem when I attempted to deal with the law of gravity within the framework of the special theory of relativity. Like most writers at the time, I tried to frame a *field-law* for gravitation, since it was no longer possible, at least in any natural way, to introduce direct action at a distance owing to the abolition of the notion of absolute simultaneity.

The simplest thing was, of course, to retain the Laplacian scalar potential of gravity, and to complete the equation of Poisson in an obvious way by a term differentiated as to time in such a way that the special theory of relativity was satisfied. The law of motion of the mass point in a gravitational field had also to be adapted to the special theory of relativity. The path was not so unmistakably marked out here, since the inert mass of a body might depend on the gravitational potential. In fact this was to be expected on account of the principle of the inertia of energy.

These investigations, however, led to a result which

raised my strong suspicions. According to classical mechanics the vertical acceleration of a body in the vertical gravitational field is independent of the horizontal component of velocity. Hence in such a gravitational field the vertical acceleration of a mechanical system or of its center of gravity works out independently of its internal kinetic energy. But in the theory I advanced the acceleration of a falling body was not independent of the horizontal velocity or the internal energy of a system.

This did not fit in with the old experimental fact that all bodies have the same acceleration in a gravitational field. This law, which may also be formulated as the law of the equality of inertial and gravitational mass, was now brought home to me in all its significance. I was in the highest degree amazed at its persistence and guessed that in it must lie the key to a deeper understanding of inertia and gravitation. I had no serious doubts about its strict validity even without knowing the results of the admirable experiments of Eötvös, which—if my memory is right—I only came to know later. I now abandoned as inadequate the attempt to treat the problem of gravitation, in the manner outlined above, within the framework of the special theory of relativity. It clearly failed to do justice to the most fundamental property of gravitation. The principle of the equality of inertial and gravitational mass could now be formulated quite clearly as follows:—In a homogeneous gravitational field all motions take place in the same way as in the absence of a gravitational field in relation to a uniformly accelerated co-ordinate system. If this principle held good for any events whatever (the "principle of equivalence"), this was an indication that the principle of relativity needed to be extended to co-ordinate systems in non-uniform motion with respect to each other, if we were to reach an easy and natural theory of the gravitational fields. Such reflections kept me busy from 1908 to 1911, and I attempted to draw special conclusions from them, of which I do not propose to speak here. For the moment the one important thing was the discovery that a reasonable theory of gravitation could only be hoped for from an extension of the principle of relativity.

What was needed, therefore, was to frame a theory whose equations kept their form in the case of non-linear transformations of the co-ordinates. Whether this was to apply to absolutely any (constant) transformations of co-ordinates or only to certain ones, I could not for the moment say.

I soon saw that bringing in non-linear transformations, as the principle of equivalence demanded, was inevitably fatal to the simple physical interpretation of the co-ordinates—i.e., that it could no longer be required that differentials of co-ordinates should signify direct results of measurement with ideal scales or clocks. I was much bothered by this piece of knowledge, for it took me a long time to see what co-ordinates in general really meant in physics. I did not find the way out of this dilemma till 1912, and then it came to me as a result of the following consideration:—

A new formulation of the law of inertia had to be found which in case of the absence of a real "gravitational field with application of an inertial system" as a co-ordinate system passed over into Galileo's formula for the principle of inertia. The latter amounts to this:—A material point, which is acted on by no force, will be represented in four-dimensional space by a straight line, that is to say by a line that is as short as possible or more correctly, an extreme line. This concept presupposes that of the length of a linear element, that is to say, a metric. In the special theory of relativity, as Minkowski had shown, this metric was a quasi-Euclidean one, i.e., the square of the "length" ds· of the linear element was a definite quadratic function of the differentials of the co-ordinates.

If other co-ordinates are introduced by means of a non-linear transformation, ds^2 remains a homogeneous function of the differentials of the co-ordinates, but the co-efficients of this function ($g_{\mu\nu}$) cease to be constant and become certain functions of the co-ordinates. In mathematical terms this means that physical (four-dimensional) space has a Riemannian metric. The time-like extremal lines of this metric furnish the law of motion of a material point which is acted on by no force apart from the forces of gravity. The co-efficients ($g_{\mu\nu}$) of this metric at the same time describe the gravitational field with reference to the co-ordinate system selected. A natural formulation of the principle of equivalence had thus been found, the extension of which to any gravitational field whatever formed a perfectly natural hypothesis.

The solution of the above-mentioned dilemma was therefore as follows:—A physical significance attaches not to the differentials of the co-ordinates but only to the Riemannian metric co-ordinated with them. A workable basis had now been found for the general theory of relativity. Two further problems remained to be solved, however.

(1) If a field-law is given in the terminology of the special theory of relativity, how can it be transferred to the case of a Riemannian metric?
(2) What are the differential laws which determine the Riemannian metric (i.e., $g_{\mu\nu}$) itself?

I worked on these problems from 1912 to 1914 together with my friend Grossmann. We found that the mathematical methods for solving problem (1) lay ready to our hands in the infinitesimal differential calculus of Ricci and Levi-Civita.

As for problem (2), its solution obviously needed invariant differential systems of the second order taken from $g_{\mu\nu}$. We soon saw that these had already been established by Riemann (the tensor of curvature). We had already considered the right field-equation for gravitation for two years before the publication of the general theory of relativity, but we were unable to see how they could be used in physics. On the contrary I felt sure that they could not do justice to experience. Moreover I believed that I could show on general considerations that a law of gravitation invariant in relation to any transformation of co-ordinates whatever was inconsistent with the principle of causation. These were errors of thought which cost me two years of excessively hard work, until I finally recognized them as such at the end of 1915 and succeeded in linking the question up with the facts of astronomical experience, after which I ruefully returned to the Riemannian curvature.

In the light of knowledge attained, the happy achievement seems almost a matter of course, and any intelligent student can grasp it without too much trouble. But the years of anxious searching in the dark, with their intense longing, their alternations of confidence and exhaustion, and the final emergence into the light;—only those who have experienced it can understand that.

THE CAUSE OF THE FORMATION OF MEANDERS IN THE COURSES OF RIVERS AND OF THE SO-CALLED BEER'S LAW

It is common knowledge that streams tend to curve in serpentine shapes instead of following the line of the maximum declivity of the ground. It is also well known to geographers that the rivers of the northern hemisphere tend to erode chiefly on the right side. The rivers of the southern hemisphere behave in the opposite manner (Beer's law). Many attempts have been made to explain this phenomenon, and I am not sure whether anything I say in the following pages will be new to the expert; some of the relevant considerations are in any case known. Nevertheless, having found nobody who thoroughly understood the elementary principles involved, I think it is proper for me to give the following short qualitative exposition of them.

First of all, it is clear that the erosion must be stronger the greater the velocity of the current where it touches the bank in question, or the more steeply it falls to zero at any particular point of the confining wall. This is equally true under all circumstances, whether the erosion depends on mechanical or on physico-chemical factors (decomposition of the ground). We must concentrate our attention on the circumstances which affect the steepness with which the velocity falls at the wall.

In both cases the asymmetry in relation to the fall in velocity in question is indirectly due to the occurrence of a circular motion to which we will next direct our attention. I begin with a little experiment which anybody can easily repeat.

Imagine a flat-bottomed cup full of tea. At the bottom there are some tea leaves, which stay there because they are rather heavier than the liquid they have displaced. If the liquid is made to rotate by a spoon, the leaves will soon collect in the center of the bottom of the cup. The explanation of this phenomenon is as follows:—The rotation of the liquid causes a centrifugal force to act on it. This in itself would give rise to no change in the flow of the liquid if the latter rotated like a solid body. But in the neighborhood of the walls of the cup the liquid is restrained by friction, so that the angular velocity with which it circulates is less there than in other places near the center. In particular, the angular velocity of circulation, and therefore the centrifugal force, will be smaller near the bottom than higher up. The result of this will be a circular movement of the liquid of the type illustrated in fig. 1. which goes on increasing until, under the influence of ground friction, it becomes stationary. The tea leaves are swept into the center by the circular movement and act as proof of its existence.

The same sort of thing happens with a curving stream (fig. 2). At every section of its course, where it is bent, a centrifugal force operates in the direction of the outside of the curve (from A to B). This force is less near the bottom, where the speed of the current is reduced by friction, than higher above the bottom. This causes a circular movement of the kind illustrated in the diagram. Even where there is

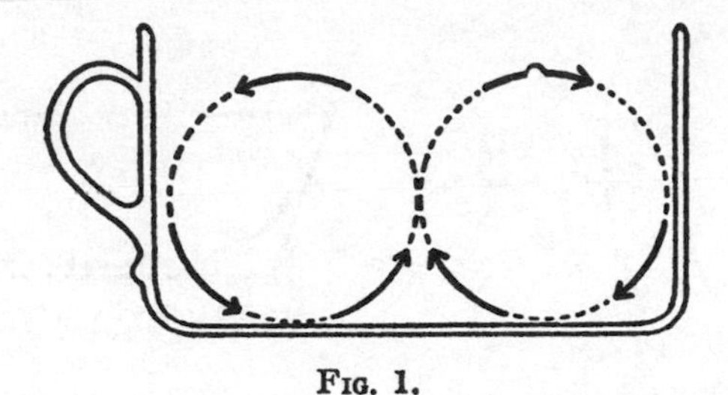

FIG. 1.

no bend in the river, a circular movement of the kind shown in fig. 2 will still take place, if only on a small scale and as a result of the earth's rotation. The latter produces a Coriolis-force, acting transversely to the direction of the current, whose right-hand horizontal component amounts to $2\,v\,\Omega \sin \phi$ per unit of mass of the liquid, where v is the velocity of the current, Ω the speed of the earth's rotation, and ϕ the geographical latitude. As ground friction causes a diminution of this force towards the bottom, this force also gives rise to a circular movement of the type indicated in fig. 2.

After this preliminary discussion we come back to

the question of the distribution of velocities over the cross section of the stream, which is the controlling factor in erosion. For this purpose we must first realize how the (turbulent) distribution of velocities takes place and is maintained. If the water which was previously at rest were suddenly set in motion by the action of an evenly diffused accelerating force, the distribution of velocities over the cross section would

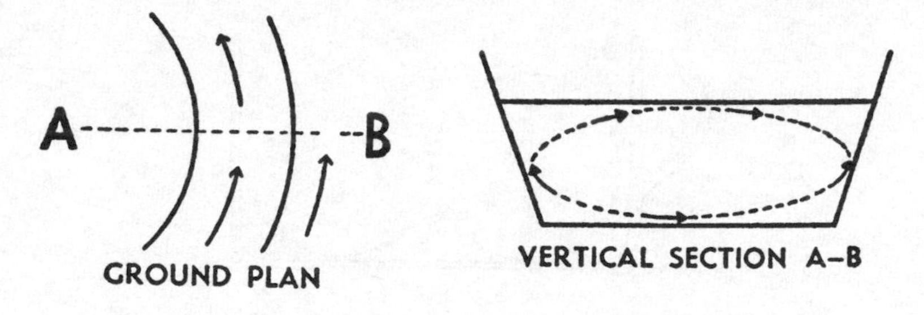

Fig. 2.

be even at first. A distribution of velocities gradually increasing from the confining walls towards the center of the cross section would only establish itself after a time, under the influence of friction at the walls. A disturbance of the (roughly speaking) stationary distribution of velocities over the cross section would only gradually set in again under the influence of fluid friction. Hydrodynamics explains the process by which this stationary distribution of velocities is established in the following way. In a systematic distribution of current (potential flow) all the vortex-filaments are concentrated at the walls.

They detach themselves and slowly move towards the center of the cross-section of the stream, distributing themselves over a layer of increasing thickness. The drop in velocity at the containing walls thereby gradually diminishes. Under the action of the internal friction of the liquid the vortex filaments in the inside of the cross section gradually get absorbed, their place being taken by new ones which form at the wall. A quasi-stationary distribution of velocities is thus produced. The important thing for us is that the adjustment of the distribution of velocities till it becomes stationary is a slow process. That is why relatively insignificant, constantly operative causes are able to exert a considerable influence on the distribution of velocities over the cross section. Let us now consider what sort of influence the circular motion due to a bend in the river or the Coriolis-force, as illustrated in fig. 2, is bound to exert on the distribution of velocities over the cross section of the river. The particles of liquid in most rapid motion will be farthest away from the walls, that is to say, in the upper part above the center of the bottom. These most rapid parts of the water will be driven by the circular motion towards the right-hand wall, while the left-hand wall gets the water which comes from the region near the bottom and has a specially low velocity. Hence in the case depicted in fig. 2 the erosion is necessarily stronger on the right side than on the left. It should be noted that this explanation is essentially based on the fact that the slow circulating movement of the water exerts a considerable influence on the distribution of velocities, because the adjustment of velocities which counteracts this consequence of the circulating movement is also a slow process on account of internal friction.

We have now revealed the causes of the formation of meanders. Certain details can, however, also be deduced without difficulty from these facts. Erosion will inevitably be comparatively extensive not

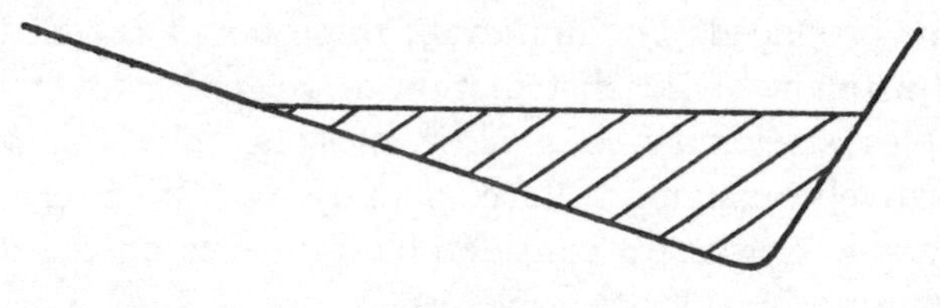

FIG. 3.

merely on the right-hand wall but also on the right half of the bottom, so that there will be a tendency to assume the shape illustrated in fig. 3.

Moreover, the water at the surface will come from the left-hand wall, and will therefore, on the left-hand side especially, be moving less rapidly than the water rather lower down. It should further be observed that the circular motion possesses inertia. The circulation will therefore only achieve its maximum extent behind the position of the greatest curvature, and the same naturally applies to the asymmetry of the erosion. Hence in the course of the erosion an advance of the wave-line of the meander-formation is bound to take place in the direction of the current. Finally, the longer the cross section of the river, the more slowly will the circular movement be absorbed by friction; the wave-line of the meander-formation will therefore increase with the cross section of the river.

THE FLETTNER SHIP

THE history of scientific and technical discovery teaches us that the human race is poor in independent thinking and creative imagination. Even when the external and scientific requirements for the birth of an idea have long been there, it generally needs an external stimulus to make it actually happen; man has, so to speak, to stumble right up against the thing before the idea comes. The Flettner ship, which is just now filling the whole world with amazement, is an excellent example of this commonplace and, for us, far from flattering truth. It also has the special attraction in its favor that the way in which the Flettner rotors work remains a mystery to most laymen, although they only involve the application of mechanical forces which every man believes himself to understand instinctively.

The scientific basis for Flettner's invention is really some two hundred years old. It has existed ever since Euler and Bernulli determined the fundamental laws of the frictionless motion of liquids. The practical possibility of achieving it, on the other hand, has only existed for a few decades—to be exact, since we have possessed usable small motors. Even then the discovery did not come automatically; chance and experience had to intervene several times first.

The Flettner ship is closely akin to the sailing ship in the way it works; as in the latter, the force of the wind is the only motive-power for propelling the ship, but instead of sails, the wind acts on vertical sheet-metal cylinders, which are kept rotating by small motors. These motors only have to overcome the small amount of friction which the cylinders encounter from the surrounding air and in their bearings. The motive power for the ship is, as I said, provided by the wind alone. The rotating cylinders look like ship's funnels, only they are several times as high and thick. The area they present to the wind is some ten times smaller than that of the equivalent tackle of a sailing ship.

"But how on earth do these rotating cylinders produce motive power?", the layman asks in despair. I will attempt to answer this question as far as it is possible to do so without using mathematical language.

In all motions of fluids (liquids or gases) where the effect of friction can be neglected, the following remarkable law holds good:—If the fluid is moving at different velocities at different points in a uniform current, the pressure is less at those points where the velocity is greater, and vice versa. This is easily understood from the primary law of the motion. If in a liquid in motion there is present a velocity with a rightward direction increasing from left to right, the individual particle of liquid is bound to undergo acceleration on its journey from left to right. In order that this acceleration may take place, a force has to act on the particle in a rightward direction. This requires that the pressure on its left edge should be stronger than that on its right. Therefore, the pressure in the liquid is greater on the left than on the right when the velocity is greater on the right than on the left.

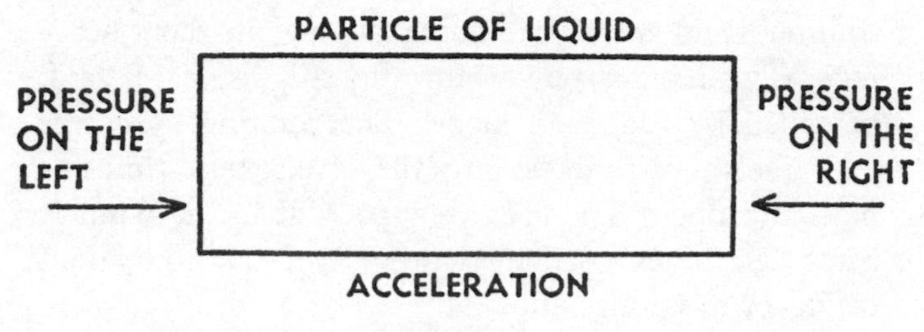

FIG. 1.

This law of the inverse ratio of the pressure to the velocity obviously makes it possible to determine the force of pressure produced by the motion of a liquid (or gas), simply by knowing the distribution of velocities in the liquid. I will now proceed to show, by a familiar example—that of the scent-spray—how the principle can be applied.

Through a pipe slightly widened at its orifice, A, air is expelled at a high velocity by means of a compressible rubber bulb. The jet of air goes on spreading uniformly in all directions as it travels, in the course of which the velocity of the current gradually sinks to zero. According to our law it is clear that there is less pressure at A, owing to the high velocity, than at a greater distance from the aperture; at A there is suction, in contrast to the more distant, stationary air. If a pipe, R, with both its ends open, is stood up with its upper end in the zone of high velocity and its lower end in a vessel filled with a liquid, the vacuum at A will draw the liquid upwards out of the vessel, and the liquid on emerging

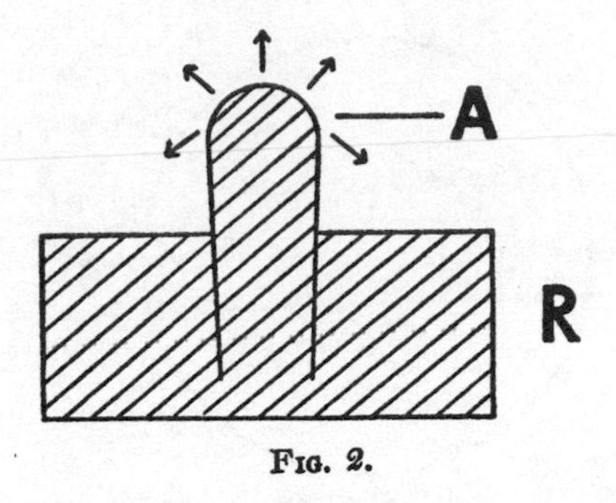

FIG. 2.

at A will be divided into tiny drops and whisked off by the current of air.

After this preliminary canter let us consider the liquid motion in a Flettner cylinder. Let C be the cylinder as seen from above. Let it not rotate to begin with. Let the wind be blowing in the direction indicated by the arrows. It has to make a certain detour round the cylinder C, in the course of which it passes A and B at the same velocity. Hence the pressure will be the same at A and B and there is no dynamic effect on the cylinder. Now let the cylinder rotate in the direction of the arrow P. The result is that the current of wind as it goes past the cylinder is divided unequally between the two sides: for the motion of the wind will be aided by the rotation of the cylinder at B, and hindered at A. The rotation of the cylinders gives rise to a motion with a greater velocity at B than at A. Hence the pressure at A is greater than at B, and the cylinder is acted upon by a force from left to right, which is made use of to propel the ship.

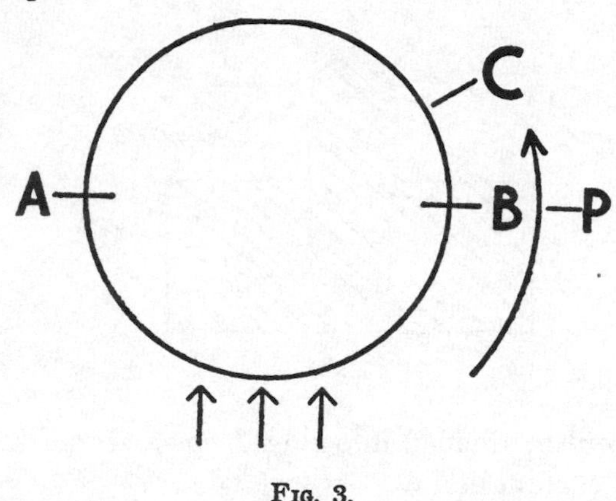

FIG. 3.

One would have thought that an inventive brain might have hit upon this idea by itself, i.e., without an extraneous cause. This, however, is what actually happened. It was observed in the course of experience that even in the absence of wind the trajectories of cannon balls exhibited considerable, irregular varying lateral deflections from the vertical plane through the initial direction of the shots. This strange phenomenon was necessarily connected, on grounds of geometry, with the rotation of the cannon balls, as there could be no other conceivable reason for a lateral asymmetry in the resistance of the air. After this phenomenon had caused a good deal of trouble to the experts, the Berlin professor of physics, Magnus, discovered the right explanation about half way through last century. It is the same as the one I have already given for the force which acts on the Flettner cylinder in the wind; only the place of the cylinder C is taken by a cannon ball rotating about the vertical axis, and that of the wind by the relative motion of the air with reference to the flying cannon ball. Magnus confirmed his explanation by experiments with a rotating cylinder which was not materially different from a Flettner cylinder. A little later the great English physicist, Lord Rayleigh, independently discovered the same phenomenon again in regard to tennis balls and also gave the correct explanation. Quite a short time ago the well known professor Prandtl has made an accurate experimental and theoretical study of fluid motion around Magnus cylinders, in the course of which he devised and carried out practically the whole of Flettner's invention. It was seeing Prandtl's experiments that put the idea into Flettner's head that this device might be used to take the place of sails. Who knows if anyone else would have thought of it if he had not?

RELATIVITY AND THE ETHER

WHY is it that alongside of the notion derived by abstraction from everyday life, of ponderable matter the physicists set the notion of the existence of another sort of matter, the ether? The reason lies no doubt in those phenomena which gave rise to the theory of forces acting at a distance, and in those properties of light which led to the wave-theory. Let us shortly consider these two things.

Non-physical thought knows nothing of forces acting at a distance. When we try to subject our experiences of bodies by a complete causal scheme, there seems at first sight to be no reciprocal interaction except what is produced by means of immediate contact, e.g., the transmission of motion by impact, pressure or pull, heating or inducing combustion by means of a flame, etc. To be sure, gravity, that is to say, a force acting at a distance, does play an important part in every day experience. But since the gravity of bodies presents itself to us in common life as something constant, dependent on no variable temporal or spatial cause, we do not ordinarily think of any cause in connection with it and thus are not conscious of its character as a force acting at a distance. It was not till Newton's theory of gravitation that a cause was assigned to it; it was then explained as a force acting at a distance, due to mass. Newton's theory certainly marks the greatest step ever taken in linking up natural phenomena causally. And yet his contemporaries were by no means satisfied with it, because it seemed to contradict the principle derived from the rest of experience that reciprocal action only takes place through direct contact, not by direct action at a distance, without any means of transmission.

Man's thirst for knowledge only acquiesces in such a dualism reluctantly. How could unity in our conception of natural forces be saved? People could either attempt to treat the forces which appear to us to act by contact as acting at a distance, though only making themselves felt at very small distances; this was the way generally chosen by Newton's successors, who were completely under the spell of his teaching. Or they could take the line that Newton's forces acting at a distance only *appeared* to act thus directly; that they were really transmitted by a medium which permeated space, either by motions or by an elastic deformation of this medium. Thus the desire for unity in our view of the nature of these forces led to the hypothesis of the ether. It certainly led to no advance in the theory of gravitation or in physics generally to begin with, so that people got into the habit of treating Newton's law of force as an irreducible axiom. But the ether hypothesis was bound always to play a part, even if it was mostly a latent one at first in the thinking of physicists.

When the extensive similarity which exists between the properties of light and those of the elastic waves in ponderable bodies was revealed in the first half of the nineteenth century, the ether hypothesis acquired a new support. It seemed beyond a doubt that light was to be explained as the vibration of an elastic, inert medium filling the whole of space. It also seemed to follow necessarily from the polarizability of light that this medium, the ether, must be of the nature of a solid body, because transverse waves are only possible in such a body and not in a fluid. This inevitably led to the theory of the 'quasi-rigid' luminiferous ether, whose parts are incapable of any motion with respect to each other beyond the small deformations which correspond to the waves of light.

This theory, also called the theory of the stationary luminiferous ether, derived strong support from the experiments, of fundamental importance for the special theory of relativity too, of Fizeau, which proved conclusively that the luminiferous ether does not participate in the motions of bodies. The phenomenon of aberration also lent support to the theory of the quasi-rigid ether.

The evolution of electrical theory along the lines laid down by Clerk Maxwell and Lorentz gave a most peculiar and unexpected turn to the development of our ideas about the ether. For Clerk Maxwell himself the ether was still an entity with purely mechanical properties, though of a far more complicated kind than those of tangible solid bodies. But neither Maxwell nor his successors succeeded in thinking out a mechanical model for the ether capable of providing a satisfactory mechanical interpretation of Maxwell's laws of the electro-dynamic field. The laws were clear and simple, the mechanical interpretations clumsy and contradictory. Almost imperceptibly theoretical physicists adapted themselves to this state of affairs, which was a most depressing one from the point of view of their mechanistic program, especially under the influence of the electro-dynamic researches of Heinrich Hertz. Whereas they had formerly demanded of an ultimate theory that it should be based upon fundamental concepts of a purely mechanical kind (e.g., mass-densities, velocities, deformations, forces of gravitation), they gradually became accustomed to admitting strengths of electrical and magnetic fields as fundamental concepts alongside of the mechanical ones, without insisting on a mechanical interpretation of them. The purely mechanistic view of nature was thus abandoned. This change led to a dualism in the sphere of fundamental concepts which was in the long run intolerable. To escape from it people took the reverse line of trying to reduce mechanical concepts to electrical ones. The experiments with β-rays and high-velocity cathode rays did much to shake confidence in the strict validity of Newton's mechanical equations.

Heinrich Hertz took no steps towards mitigating this dualism. Matter appears as the substratum not only of velocities, kinetic energy, and mechanical forces of gravity, but also of electro-magnetic fields. Since such fields are also found in a vacuum—i.e., in the unoccupied ether—the ether also appears as the substratum of electro-magnetic fields, entirely similar in nature to ponderable matter and ranking alongside it. In the presence of matter it shares in the motions of the latter and has a velocity everywhere in empty space; the etheric velocity nowhere changes discontinuously. There is no fundamental distinction between the Hertzian ether and ponderable matter (which partly consists of the ether).

Hertz's theory not only suffered from the defect that it attributed to matter and the ether mechanical and electrical properties, with no rational connection between them; it was also inconsistent with the result of Fizeau's famous experiment on the velocity of the propagation of light in a liquid in motion and other well authenticated empirical facts.

Such was the position when H. A. Lorentz entered the field. Lorentz brought theory into harmony with experiment, and did it by a marvelous simplification of basic concepts. He achieved this advance in the science of electricity, the most important since Clerk Maxwell, by divesting the ether of its mechanical, and matter of its electro-magnetic properties. Inside material bodies no less than in empty space the ether alone, not atomically conceived matter, was the seat of electro-magnetic fields. According to Lorentz the elementary particles of matter are capable *only* of executing movements; their electromagnetic activity is entirely due to the fact that they carry electric charges. Lorentz thus succeeded in reducing all electro-magnetic phenomena to Maxwell's equations for a field in vacuo.

As regards the mechanical nature of Lorentz's ether, one might say of it, with a touch of humor, that immobility was the only mechanical property which Lorentz left it. It may be added that the whole difference which the special theory of relativity made in our conception of the ether lay in this, that it divested the ether of its last mechanical quality, namely immobility. How this is to be understood I will explain immediately.

The Maxwell-Lorentz theory of the electro-magnetic field served as the model for the space-time theory and the kinematics of the special theory of relativity. Hence it satisfies the conditions of the special theory of relativity; but looked at from the standpoint of the latter, it takes on a new aspect. If C is a co-ordinate system in respect to which the Lorentzian ether is at rest, the Maxwell-Lorentz equations hold good first of all in regard to C. According to the special theory of relativity these same equations hold good in exactly the same sense in regard to any new co-ordinate system C, which is in uniform translatory motion with respect to C. Now comes the anxious question, Why should I distinguish the system C, which is physically perfectly equivalent to the systems C′, from the latter by assuming that the ether is at rest in respect to it? Such an asymmetry of the theoretical structure, to which there is no corresponding asymmetry in the system of empirical facts, is intolerable to the theorist. In my view the physical equivalence of C and C′ with the assumption that the ether is at rest in respect to C but in motion with respect to C′, though not absolutely wrong from a logical point of view, is nevertheless unsatisfactory.

The most obvious line to adopt in the face of this situation seemed to be the following:—There is no such thing as the ether. The electro-magnetic fields are not states of a medium but independent realities, which cannot be reduced to terms of anything else and are bound to no substratum, any more than are the atoms of ponderable matter. This view is rendered the more natural by the fact that, according to Lorentz's theory, electro-magnetic radiation carries impulse and energy like ponderable matter, and that matter and radiation, according to the special theory of relativity, are both of them only particular forms of distributed energy, inasmuch as ponderable matter loses its exceptional position and merely appears as a particular form of energy.

In the meantime more exact reflection shows that this denial of the existence of the ether is not demanded by the restricted principle of relativity. We can assume the existence of an ether; but we must abstain from ascribing a definite state of motion to it, i.e., we must divest it by abstraction of the last mechanical characteristic which Lorentz left it. We shall see later on that this way of looking at it, the intellectual possibility of which I shall try to make clearer by a comparison that does not quite fit at all points, is justified by the results of the general theory of relativity.

Consider waves on the surface of water. There are two quite different things about this phenomenon which may be described. One can trace the progressive changes which take place in the undulating surface where the water and the air meet. One can also —with the aid of small floating bodies, say—trace the progressive changes in the position of the individual particles. If there were in the nature of the case no such floating bodies to aid us in tracing the movement of the particles of liquid, if nothing at all could be observed in the whole procedure except the fleeting changes in the position of the space occupied by the water, we should have no ground for supposing that the water consists of particles. But we could none the less call it a medium.

Something of the same sort confronts us in the electro-magnetic field. We may conceive the field as consisting of lines of force. If we try to think of these lines of force as something material in the ordinary sense of the word, there is a temptation to ascribe the dynamic phenomena involved to their motion, each single line being followed out through time. It is, however, well known that this way of looking at the matter leads to contradictions.

Generalizing, we must say that we can conceive of extended physical objects to which the concept of motion cannot be applied. They must not be thought of as consisting of particles, whose course can be followed out separately through time. In the language of Minkowski this is expressed as follows:—Not every extended entity in the four-dimensional world can be regarded as composed of world-lines. The special principle of relativity forbids us to regard the ether as composed of particles the movements of which can be followed out through time, but the theory is not incompatible with the ether hypothesis as such. Only we must take care not to ascribe a state of motion to the ether.

From the point of view of the special theory of relativity the ether hypothesis does certainly seem an empty one at first sight. In the equations of an electro-magnetic field, apart from the density of the electrical charge nothing appears except the strength of the field. The course of electro-magnetic events in a vacuum seems to be completely determined by that inner law, independently of other physical quantities. The electro-magnetic field seems to be the final irreducible reality, and it seems superfluous at first sight to postulate a homogeneous, isotropic etheric medium, of which these fields are to be considered as states.

On the other hand, there is an important argument in favor of the hypothesis of the ether. To deny the existence of the ether means, in the last analysis, denying all physical properties to empty space. But such a view is inconsistent with the fundamental facts of mechanics. The mechanical behavior of a corporeal system floating freely in empty space depends not only on the relative positions (intervals) and velocities of its masses, but also on its state of rotation, which cannot be regarded physically speaking as a property belonging to the system as such. In order to be able to regard the rotation of a system at least formally as something real, Newton regarded space as objective. Since he regards his absolute space as a real thing, rotation with respect to an absolute space is also something real to him. Newton could equally well have called his absolute space "the ether"; the only thing that matters is that in addition to observable objects another imperceptible entity has to be regarded as real, in order for it to be possible to regard acceleration, or rotation, as something real.

Mach did indeed try to avoid the necessity of postulating an imperceptible real entity, by substituting in mechanics a mean velocity with respect to the totality of masses in the world for acceleration with respect to absolute space. But inertial resistance with respect to the relative acceleration of distant masses presupposes direct action at a distance. Since the modern physicist does not consider himself entitled to assume that, this view brings him back to the ether, which has to act as the medium of inertial action. This conception of the ether to which Mach's approach leads, differs in important respects from that of Newton, Fresnel and Lorentz. Mach's ether not only *conditions* the behavior of inert masses but is also conditioned, as regards its state, by them.

Mach's notion finds its full development in the ether of the general theory of relativity. According to this theory the metrical properties of the space-time continuum in the neighborhood of separate space-time points are different and conjointly conditioned by matter existing outside the region in question. This spatio-temporal variability of the relations of scales and clocks to each other, or the knowledge that "empty space" is, physically speaking, neither homogeneous nor isotropic, which compels us to describe its state by means of ten functions, the gravitational potentials $g_{\mu\nu}$, has no doubt finally disposed of the notion that space is physically empty. But this has also once more given the ether notion a definite content—though one very different from that of the ether of the mechanical wave-theory of light. The ether of the general theory of relativity is a medium which is itself free of *all* mechanical and kinematic properties, but helps to determine mechanical (and electro-magnetic) happenings.

The radical novelty in the ether of the general theory of relativity as against the ether of Lorentz lies in this, that the state of the former at every point is determined by the laws of its relationship with matter and with the state of the ether at neighboring points expressed in the form of differential equations, whereas the state of Lorentz's ether in the absence of electro-magnetic fields is determined by nothing outside it and is the same everywhere. The ether of the general theory of relativity can be transformed intellectually into Lorentz's through the substitution of constants for the spatial functions which describe its state, thus neglecting the causes conditioning the latter. One may therefore say that the ether of the general theory of relativity is derived through relativization from the ether of Lorentz.

The part which the new ether is destined to play in the physical scheme of the future is still a matter of uncertainty. We know that it determines both material relations in the space-time continuum, e.g., the possible configurations of solid bodies, and also gravitational fields; but we do not know whether it plays a material part in the structure of the electric particles of which matter is made up. Nor do we know whether its structure only differs materially from that of Lorentz's in the proximity of ponderable masses, whether, in fact, the geometry of spaces of cosmic extent is, taken as a whole, almost Euclidean. We can, however, maintain on the strength of the relativistic equations of gravitation that spaces of cosmic proportions must depart from Euclidean behavior if there is a positive mean density of matter, however small, in the Universe. In this case the Universe must necessarily form a closed space of finite size, this size being determined by the value of the mean density of matter.

If we consider the gravitational field and the electro-magnetic field from the standpoint of the ether hypothesis, we find a notable fundamental difference between the two. No space and no portion of space without gravitational potentials; for these give it its metrical properties without which it is not thinkable at all. The existence of the gravitational field is directly bound up with the existence of space.

On the other hand a portion of space without an electro-magnetic field is perfectly conceivable, hence the electro-magnetic field, in contrast to the gravitational field, seems in a sense to be connected with the ether only in a secondary way, inasmuch as the formal nature of the electro-magnetic field is by no means determined by the gravitational ether. In the present state of theory it looks as if the electro-magnetic field, as compared with the gravitational field, were based on a completely new formal motive; as if nature, instead of endowing the gravitational ether with fields of the electro-magnetic type, might equally well have endowed it with fields of a quite different type, for example, fields with a scalar potential.

Since according to our present-day notions the primary particles of matter are also, at bottom, nothing but condensations of the electro-magnetic field, our modern schema of the cosmos recognizes two realities which are conceptually quite independent of each other even though they may be causally connected, namely the gravitational ether and the electro-magnetic field, or—as one might call them—space and matter.

It would, of course, be a great step forward if we succeeded in combining the gravitational field and the electro-magnetic field into a single structure. Only so could the era in theoretical physics inaugurated by Faraday and Clerk Maxwell be brought to a satisfactory close.

The antithesis of ether and matter would then fade away, and the whole of physics would become a completely enclosed intellectual system, like geometry, kinematics and the theory of gravitation, through the general theory of relativity. An exceedingly brilliant attempt in this direction has been made by the mathematician H. Weyl; but I do not think that it will stand in the face of reality. Moreover, in thinking about the immediate future of theoretical physics we cannot unconditionally dismiss the possibility that the facts summarized in the quantum theory may set impassable limits to the field theory.

We may sum up as follows: According to the general theory of relativity space is endowed with physical qualities; in this sense, therefore, an ether exists. In accordance with the general theory of relativity space without an ether is inconceivable. For in such a space there would not only be no propagation of light, but no possibility of the existence of scales and clocks, and therefore no spatio-temporal distances in the physical sense. But this ether must not be thought of as endowed with the properties characteristic of ponderable media, as composed of particles the motion of which can be followed; nor may the concept of motion be applied to it.

ADDRESS AT COLUMBIA UNIVERSITY, NEW YORK, JANUARY 15

SCIENCE as something existing and complete is the most objective thing known to man. But science in the making, science as an end to be pursued, is as subjective and psychologically conditioned as any other branch of human endeavor—so much so that the question, What is the purpose and meaning of science? receives quite different answers at different times and from different sorts of people.

It is, of course, universally agreed that science has to establish connections between the facts of experience, of such a kind that we can predict further occurrences from those already experienced. Indeed, according to the opinion of many positivists the completest possible accomplishment of this task is the only end of science.

I do not believe, however, that so elementary an ideal could do much to kindle the investigator's passion from which really great achievements have arisen. Behind the tireless efforts of the investigator there lurks a stronger, more mysterious drive: it is existence and reality that one wishes to comprehend. But one shrinks from the use of such words, for one soon gets into difficulties when one has to explain what is really meant by "reality" and by "comprehend" in such a general statement.

When we strip the statement of its mystical elements we mean that we are seeking for the simplest possible system of thought which will bind together the observed facts. By the "simplest" system we do not mean the one which the student will have the least trouble in assimilating, but the one which contains the fewest possible mutually independent postulates or axioms; since the content of these logical, mutually independent axioms represents that remainder which is not comprehended.

When a man is talking about scientific subjects, the little word "I" should play no part in his expositions. But when he is talking about the purposes and aims of science, he should be permitted to speak of himself; for a man experiences no aims and desires so immediately as his own. The special aim which I have constantly kept before me is logical unification in the field of physics. To start with, it disturbed me that electro-dynamics should pick out *one* state of motion in preference to others, without any experimental justification for this preferential treatment. Thus arose the special theory of relativity, which, moreover, welded together into comprehensible unities the electrical and magnetic fields, as well as mass and energy, or momentum and energy, as the case may be. Then out of the endeavor to understand inertia and gravitation as having a unified character there arose the general theory of relativity, which also avoided those implicit axioms which underlie our thinking when we use special co-ordinate systems in the process of formulating basic laws.

At the present time it is particularly disturbing that the gravitational field and the electrical field should enter into the theory as mutually independent fundamental concepts. After many years of effort, however, an appropriate logical unification has been achieved—so I believe—through a new mathematical method, which I have invented together with my distinguished collaborator, Dr. W. Mayer.

In the meantime there still remains outstanding an important problem of the same kind, which has often been proposed but has so far found no satisfactory solution—namely the explanation of atomic structure in terms of field theory. All of these endeavors are based on the belief that existence should have a completely harmonious structure. Today we have less ground than ever before for allowing ourselves to be forced away from this wonderful belief.